Overdentures in
General Dental Practice

Overdentures in General Dental Practice

R. M. Basker, DDS, MGDS, LDS
A. Harrison,* TD, PhD, BDS, FDS
J. P. Ralph, DDS, FDS, HDD
C. J. Watson, PhD, BDS, FDS

*Division of Restorative Dentistry,
Leeds Dental Institute, University of Leeds*
**Department of Prosthodontics and Periodontology,
University of Bristol*

Third Edition

Published by the British Dental Association
64 Wimpole Street, London W1M 8AL

First edition 1983
Second edition 1988

ISBN 0 904588 42 4

Typeset by Latimer Trend Ltd, Plymouth; printed and bound in
Great Britain by Eyre & Spottiswoode Ltd, Margate

Contents

Foreword to the First Edition

This book contains the six articles which were published as a series in the *British Dental Journal*, Volumes 154 and 155, from May to July, 1983. The intention of the authors was to present the current viewpoint on overdentures — a concept which has attracted an increasing amount of interest in the last few years. Rather than punctuating the narrative with references, a bibliography of selected readings has been included after the final chapter.

Foreword

The last 10 years has seen a rapid increase in the use of dental implants as an aid in the treatment of tooth loss. During this time, clinical trials have shown dental implants to be predictable and to have a high long-term survival rate. We have therefore taken the opportunity of adding three new chapters on the use of implants as overdenture abutments in the edentulous patient. We have also revised and updated the other six chapters. We have continued our policy of not including references in the text as a consequence of a favourable critical comment.

Acknowledgements

We are most grateful to Miss Anna M. Durbin, Mr Angus J. Robertson and Mr David Hawkridge, all of the Medical and Dental Illustrations Unit in the University of Leeds, for preparing the illustrative material. Our thanks are extended to Mr Brian Oliver, Mr Jack E. Pentland, Mr Barry S. Dransfield and Mr Robert Ruddy for their technical expertise, to Miss Alison Peacock for her expert secretarial help, and to Mr Leslie Smillie for his editorial skills.

1

The Evidence for Overdentures

An overdenture is a prosthesis that derives support from one or more abutment teeth by completely enclosing them beneath its impression surface (figs 1.1 and 1.2). The use of roots to support a denture is not a new idea, some of the earliest reports of the technique having been written in the middle of the nineteenth century. Other references appeared in British and American books and journals around the turn of the century. There followed a lull, until interest in the concept was revived in the 1950s by publications from Scandinavia and the USA. Little attention was paid to overdentures in the UK until about 10 years later, since when there has been more widespread acceptance.

The overdenture procedure harnesses those skills which the clinician already possesses, namely, the understanding and practice of periodontal, endodontic, and general prosthetic techniques.

This first chapter is not intended as a comprehensive review of the enormous amount of literature available, rather, it is intended to give the reader an idea of the basic premises on which overdenture treatment is based.

Maintenance of alveolar bone
Bone is constantly being remodelled by processes of resorption and apposition. Following loss of the teeth, resorption of the alveolar ridges occurs, the rate varying between different individuals and within the same individual at different times.

Considerable research efforts have gone into studying the patterns of bone resorption following tooth extraction, particularly in the mandible, and it is quite clear that the removal of all natural teeth and the wearing of complete dentures results in gradual loss of alveolar bone. For example, it was observed in one study that the average reduction of anterior mandibular ridge height was 9–10 mm over a 25-year period,

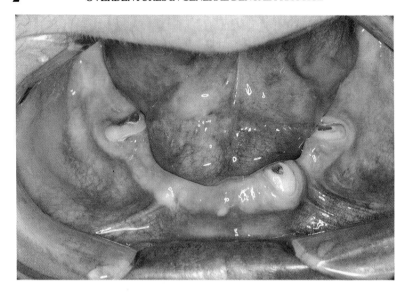

Fig. 1.1 Retention of $\overline{6|35}$ as overdenture abutments.

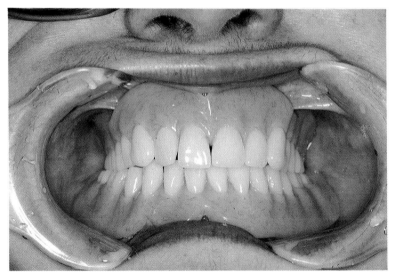

Fig. 1.2 Conventional upper denture and lower overdenture.

whereas only 2.5–3 mm were lost from the maxillary ridge. This finding was in close agreement with earlier work which showed that maxillary and mandibular alveolar resorption rates over a 7-year period were in

Fig. 1.3 The dotted line represents the height of the bone after wearing dentures for 5 years. (a) Retention of 3̅|3̅ with overdenture. (b) Conventional denture. (After Crum J, Rooney G E. Alveolar bone loss in overdentures: a five year study. *J Prosthet Dent* 1978; **40:** 610–613.)

the ratio 1:4. Thus the mandibular ridge is particularly susceptible to change.

Another study was performed on two groups of men who still had their natural teeth prior to prosthetic treatment. The first group was treated with a complete upper denture and a lower overdenture, retaining the two canine roots. Patients in the second group were provided with conventional complete upper and lower dentures. Cephalometric radiographs were taken over a 5-year period to assess the amount of bone resorption in the vertical plane. In the overdenture group an average of only 0.6 mm of vertical height was lost from the anterior part of the mandible, whilst in the conventional denture group there was a loss of 5.2 mm; bone loss from the maxilla was similar in both groups. These results show that the roots help to maintain bone, not only in their immediate vicinity, but also in adjacent areas (fig. 1.3). It has been suggested that the discrete proprioceptive ability within the periodontal ligaments of teeth under an overdenture acts as a signal against physiological overload of the system, thus reducing bone resorption.

More patients experience difficulty with complete lower than with complete upper dentures. Denture problems appear to intensify with age, and the number of patients with these problems may well increase considering that there are now approximately 8.5 million people in the UK over the age of 65 years. Because many of the problems associated with lower complete dentures can be attributed to the atrophied mandibular ridge, alveolar bone must be preserved if one wishes to ensure denture comfort and stability into old age. However, the upper ridge is also susceptible to excessive resorption if an upper complete denture is opposed by natural teeth. The unfavourable loading may lead to the denture-bearing tissues being replaced by a fibrous anterior ridge (bearing in mind that the lower anterior teeth are commonly the last to be extracted). This condition is probably quite common since approximately 30% of all partially edentulous adults wear a complete upper denture opposed by natural teeth with or without a partial denture. Over the years many authors have presented different approaches to the preservation of the residual ridge, including the use of non-anatomical tooth design, dietary control and special impression techniques. However, these approaches deal with the problem after the teeth have been extracted and major bone loss initiated, and can thus have only minimal long-term benefit.

We know from clinical experience, however, that alveolar bone is maintained whilst healthy teeth or roots remain. Figure 1.4 illustrates clearly that whilst considerable resorption has occurred in the molar regions as a result of tooth extraction many years ago, buttresses of bone have been maintained around the overdenture abutments and the presence of these teeth has helped to preserve the bone in the incisor region. The influence on ridge shape of retaining teeth and roots is also shown in figure 1.5.

The presence of overdenture abutments permits the stresses of occlusion to be spread over a larger area as the support of the periodontal ligament is brought into function. Thus, the retention of a number of these abutments maintains a positive ridge form, with a greater height and volume of alveolar bone, and the denture is therefore more stable. Retention should also be improved because of the greater supporting area available.

It is reasonable to assume that the likelihood of traumatic loading of the denture-bearing mucosa will be reduced. The maintenance of alveolar bone around the retained roots is, of course, dependent upon the continued health of the periodontal tissues.

Sensory feedback

Whereas the maintenance of alveolar bone around overdenture abutments is an advantage which is both tangible and well documented, the benefits of preserving sensory feedback are less obvious, harder to investigate and thus, at the moment, open to conjecture.

The success or failure of a prosthesis is dependent upon the integration of sensory feedback and motor response, the effective functioning of the masticatory system relying on feedback from receptors in the temporomandibular joints, muscles, ligaments, oral mucosa, teeth and periodontal tissues. The extraction of the teeth results in the loss of

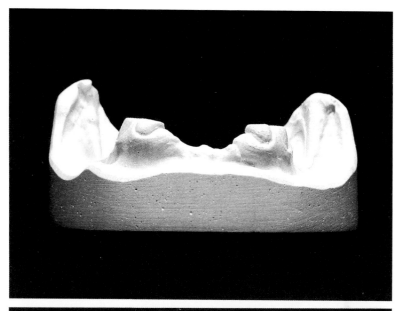

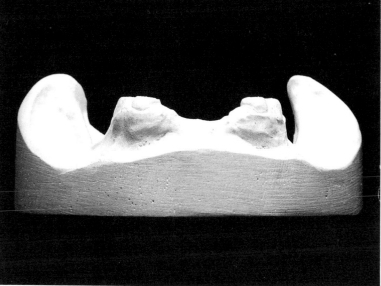

Fig. 1.4 The cast at the bottom of the figure was obtained shortly after the immediate restoration of the incisors. The benefit to the shape of the ridge of keeping abutment teeth is evident when comparing with the cast at the top of the figure which was obtained 5 years later. (Reproduced from the 1988 *Dental Annual* with the permission of the publisher, Butterworth-Heinemann.)

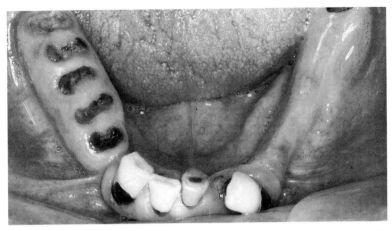

Fig. 1.5 Although the mouth is neglected, the retained roots on the patient's right-hand side have maintained the width of the ridge.

discrete proprioception that has been part of the sensory programme throughout life. The edentulous ridges cannot give the same sensory feedback or physiological support.

The belief that sensory input from the periodontal ligament is more influential than that from the dental pulp is substantiated by animal research which has shown that removal of pulpal tissue makes no change to the response. Furthermore, it has been shown in humans that vital and pulpless teeth were equally capable of detecting minimal loads.

The likely influences of sensory feedback from the teeth are: (a) assistance in controlling masticatory force; (b) assistance in recognising the size and texture of objects placed between the teeth; (c) assistance in monitoring the position of the mandible during function.

Many studies of sensory feedback have been conducted over the years and, in general terms, they fall into two main groups: those concerned with recognition of load thresholds and those investigating tactile sensitivity, that is, the ability to differentiate the sizes of objects placed between the teeth.

Minimal load thresholds
Within the periodontal ligament are receptors capable of highly sensitive load discrimination. Animal research has shown that the periodontal ligaments of the canine teeth are more richly represented by mechanoreceptors than any other teeth in that same species; these teeth are thus more responsive to stimulation. Human studies have shown

there to be little difference between the incisors and canines in their ability to discriminate between varying levels of force.

Several workers have shown that, in humans, the anterior natural teeth can detect a load of as little as 1 g, whereas the posterior teeth have a minimum threshold of 8-10 g. In contrast, one investigation reported that the minimum threshold for complete denture wearers was approximately 125 g when measured in the first premolar region.

Studies of the appreciation of the direction of the load have reached differing conclusions. On the one hand there is comment that the sensitivity to lateral load is 2-5 times greater than to axial load; in another investigation the subjects were unable to discriminate between forces applied along the long axis of the crown and forces applied to the labial surface and directed at right angles to the long axis of the crown. Although individual receptor response is related to the direction of the stimulus applied to the tooth, the presence of a group of receptors means that there will always be individual receptors capable of responding to load, regardless of the direction from which it is applied.

On balance, current evidence suggests that the teeth are more sensitive to lateral forces than to axial forces. This finding may be seen as an advantage as it can be argued that the lower threshold to lateral force is a protective mechanism which limits this more damaging force and thus helps to prevent damage to the periodontal tissues.

It seems that wearers of conventional dentures have a higher minimum load threshold. Thus, the protective mechanism which guards the supporting tissues is likely to be less precise. The presence of overdenture abutments may enable the denture wearer to apply higher occlusal forces during mastication and offers improved discriminatory ability at both higher and lower force levels. It is likely that the overdenture wearer is more efficient at controlling and grading the higher forces needed to break up a mouthful of food. This may be because the higher force seats the denture firmly onto the abutment teeth and enhances the performance of the sensory receptors in the periodontal ligaments.

Tactile sensitivity discrimination

Fine wires or foils have commonly been used in studies to determine the ability of subjects to discriminate differences in thickness of objects placed between natural and/or artificial teeth. One study concluded that complete denture wearers were approximately six times less efficient than dentate persons at detecting very small objects between the teeth. Patients wearing overdentures were more able to discriminate objects of different sizes placed between the teeth than were patients wearing conventional complete dentures. In contrast, another study reported no significant differences between three groups of subjects — those with natural teeth, those with overdentures, and those wearing conventional complete dentures; it was argued that the loss of periodontal receptors was of no crucial importance to discriminatory ability since it was likely that muscle spindles played the major role.

In comparison with dentate subjects, denture wearers are less able to discriminate fine particles or changes in food texture. Unstable dentures further reduce this lowered discriminatory sense for fine particles, although this sense does seem to improve as the wearer gains functional adaptation during the life of the prosthesis. It has also been shown that instability of conventional dentures reduces the ability of patients to perceive changes in the hardness of objects placed between the teeth. One would expect the increased stability of the overdenture to improve these same aspects of sensory discrimination, and to provide the denture wearer with better discrimination of the size and texture of food particles.

Masticatory performance
It is likely that the increased sensory feedback inherent in overdenture patients gives them a functional advantage over conventional complete denture patients. One study recorded the results of a masticatory performance test in which each patient chewed a 3 g portion of carrot for 40 masticatory strokes. The bolus was then expectorated and the percentage of chewed food passing through a No. 12 sieve was then recorded as follows: natural dentition (90%); complete dentures (59%); overdentures (79%). This supports the contention that overdenture patients are able to chew their food more effectively. There is also evidence that oral function in subjects with overdentures supported by osseointegrated implants is improved. A recent study demonstrated an increase in functional biting forces and chewing efficiency when subjects were assessed before and after osseointegrated implants were placed. The evaluation comprised a subjective and clinical examination.

In summary it would appear reasonable to make the following points:
1. The periodontal ligament is only one of the sites of sensory feedback.
2. If teeth are retained as overdenture abutments, the subsequent denture is likely to be more stable.
3. When the stable denture is in contact with the overdenture abutments, there will be improved tactile sense and neuromuscular performance.

4. The overdenture patient is able to exert higher forces during mastication and has more precise appreciation and control of the forces.

Reduction of psychological trauma

The loss of all the remaining teeth can be a disturbing emotional experience for many people and it has been suggested that such a disturbance may precipitate long-term denture problems. It can be argued that the retention of overdenture abutments prevents this feeling of total loss and perhaps makes the patient more readily able to adjust to the transition to complete dentures. Evidence about the attitude of different groups of patients to the prospect of losing their remaining teeth and having complete dentures was presented in a study of adult dental health in the UK in 1988. Not surprisingly, over 60% of those who did not wear dentures found the thought of complete dentures a very upsetting one; nearly half of those who thought they were likely to need complete dentures eventually had a similar degree of concern. Even amongst those who already wore dentures in conjunction with their remaining natural teeth nearly 40% had the same degree of worry; the level of concern was greater amongst those who experienced problems with the existing dentures.

Although there is no firm evidence to show that the retention of overdenture abutments eliminates these concerns and smooths the way towards successful complete denture wearing, it can surely be safely assumed that the functional benefits which come from the preservation of a well-formed ridge can only be of assistance to the apprehensive patient who is faced with making the transition from what remains of the natural dentition to the totally artificial one.

Conclusions

The retention of healthy abutment teeth maintains alveolar bone, enhances sensory feedback, improves masticatory performance and may, in a small number of cases, help the patient to adjust mentally to dentures.

The first two advantages confer on the overdenture improved stability and retention, create more favourable loading of the denture-bearing tissues and promote more accurate control of the appliances during function.

In addition to these inherent benefits, the overdenture can be of particular help in dealing with certain clinical problems. Inevitably, there are also drawbacks to the technique. The clinical advantages and disadvantages will be emphasised in later chapters.

2
Indications for Overdentures:
Patient Selection

In Chapter 1, the reasons for retaining teeth to serve as abutments for overdentures were presented. In this chapter we will look at the clinical situations in which this form of treatment is of particular benefit to patients.

Indications for overdentures

The possibility of retaining a number of teeth to serve as overdenture abutments should be considered on each occasion that the decision to extract teeth and provide dentures is made. In this way the overdenture will gradually become more widely accepted as part of routine prosthetic treatment, and the part which it can play in stabilising the extensive partial denture, or in easing the transition to complete dentures, will be more fully appreciated. In addition, there are a number of specific clinical situations in which the concept is of particular value.

The single complete denture

There is no doubt that patients often experience considerable difficulty if they are compelled to wear a complete denture in one jaw opposed by an intact or substantially intact natural dentition in the other (fig. 2.1). The alignment of the occlusal plane in the natural dentition makes it difficult to develop satisfactory occlusal balance and the complete denture is often unfavourably loaded during excursive movements of the mandible. In addition, such patients generally exert greater masticatory forces than those who wear complete dentures in both jaws, and as a consequence the mucosa underlying the complete denture tends to be loaded excessively and alveolar resorption proceeds more rapidly. Many of the patients who wear this potentially damaging combination of denture opposed by natural teeth are only middle-aged and can realistically expect to live for a further 30 or more years, during which

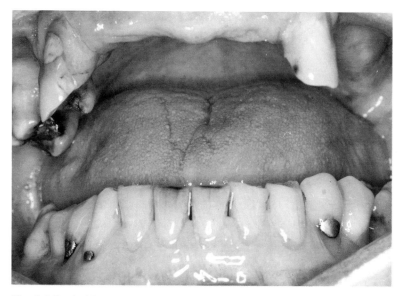

Fig. 2.1 Depleted maxillary dentition. It is proposed that 3|3 will be retained as root abutments to support a complete overdenture against an almost intact lower arch.

time enormous damage can be done to the denture-bearing tissues. Common complaints include denture looseness, denture fracture, sore gums and a deteriorating appearance as the bone is gradually lost and the denture is driven further into the tissues by the opposing natural teeth.

These considerable problems can be prevented by preserving abutment teeth and with an overdenture it is possible to achieve a more uniform distribution of load on to the supporting structures. The presence of the abutments also helps to slow down the rate of alveolar resorption, thus improving the prospects of long-term prosthetic success. Where the combination of an upper edentulous arch opposed by natural anterior teeth exists, together with evidence of tissue damage, it may be necessary to consider more drastic treatment of the lower jaw. Extraction of the remaining teeth and the provision of a complete lower denture is an irreversible step which is fraught with danger. On the other hand the preservation of overdenture abutments in order to maintain a good ridge shape is a line of treatment which should not be overlooked. This approach allows a balanced occlusion to be created and thus helps to reduce the damaging forces.

Having made all these points, it must be said that many of these patients have reached the stage where a decision has had to be made on

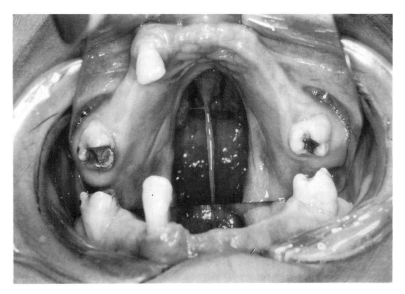

Fig. 2.2 An extensive cleft of the soft and hard palates with only 73|7 remaining in the upper jaw to serve as abutments for an overdenture.

whether or not to extract the remaining teeth in one or both jaws largely as a result either of neglect or an unwillingness to seek regular treatment and maintenance; they must therefore be regarded as patients who are at a higher risk as far as further dental disease is concerned. Unless a change in attitude can be created they may not be candidates for overdenture treatment for reasons which are presented later in this chapter.

Cleft palates and surgical defects
The jaw structure and dental arches may be deficient because of some disturbance in the normal growth and development, the most common of which is the cleft palate (fig. 2.2). The removal of tumours from the maxillary sinus or from the floor of the mouth may also leave large defects which require prosthetic reconstruction. Occasionally, a similar problem may arise following gun-shot wounds to the facial region.

The presence of such congenital or acquired defects can pose considerable problems in the design and construction of dentures, either partial or complete. In some instances the defect may be so extensive, or the tissue morphology so unfavourable, that there would be no prospect of providing a satisfactory denture by conventional methods. In these circumstances it often becomes critical to select suitable teeth which can be used as overdenture abutments. If there is disparity in

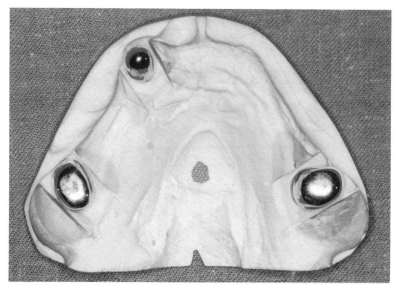

Fig. 2.3 Gold copings prepared for 73|7 abutments.

maxillomandibular arch form, it is often helpful to maintain the crown structure of the teeth, suitably modified, rather than to reduce them to root abutments (fig. 2.3). Where retention is likely to be a problem it may also be possible to use some form of precision attachment.

Hypodontia

When only a few units of the permanent dentition fail to develop, it is often possible to effect a satisfactory restoration by means of bridge work or conventional partial dentures. When substantial numbers of teeth are absent and the remaining teeth are irregularly positioned (fig. 2.4), or when there is also an associated disparity in growth of the dental arches, the best form of restoration may often be an overdenture, either partial or complete. If in addition there is a need to restore facial height this will often be achieved more easily if most of the coronal tooth structure is maintained as the overdenture abutment. A modified crown preparation may be undertaken and a gold coping fitted to protect the underlying tooth structure from caries.

Severe tooth wear

The increased prevalence of this condition, an example of which is shown in figure 2.5, has prompted a number of investigations into the

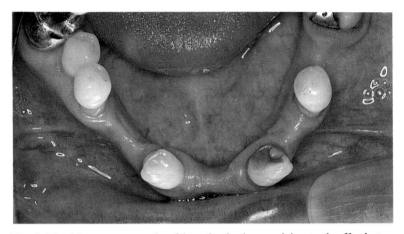

Fig. 2.4 In this severe example of hypodontia the remaining teeth offer better support for an overdenture than for either bridge work or for a conventional partial denture.

aetiology. The consensus is that in most cases more than one cause is responsible for the wear that has occurred. Commonly, attrition and erosion go hand-in-hand (Fig. 2.6); erosion of the enamel may expose the underlying dentine to accelerated attrition or alternatively attrition of incisal or occlusal enamel may potentiate the action of erosive agents

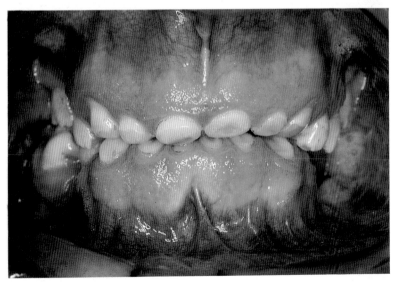

Fig. 2.5 An example of severe tooth wear of upper and lower teeth.

Table 2.1 Some causes of severe tooth wear

Erosion
    Diet
        excessive consumption of fruit (citrus and others)
        excessive consumption of drinks (citrus, cola-type, carbonated, lemon tea)
        medication (HCl, Vitamin C tablets)
    Intrinsic
        anorexia nervosa
        bulimia
        alcoholism
        hiatus hernia
    Industrial — acids
(The action of all the above can be potentiated by lack of saliva and consequent reduction in buffering capacity)
Abrasion — abrasive foods, brushing habits
Attrition — parafunction, partial loss of teeth

producing the characteristic cupping of the occlusal surface. Abrasion is perhaps a less common cause.

Faced with a large number of possible causative factors, as listed in Table 2.1, it is clearly important to attempt to establish a diagnosis in order to eliminate potent damaging factors before embarking on any

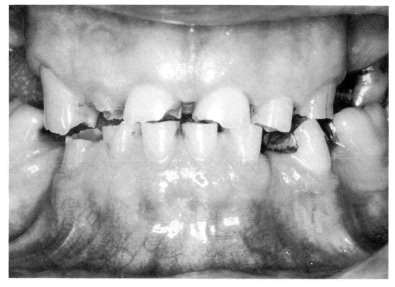

Fig. 2.6 Evidence of attrition and erosion affecting upper and lower anterior teeth.

restorative procedure. For example, if a dietary factor predominates it would be folly to cover root faces with an overdenture whilst there is still a risk of an acidic, erosive liquid being constantly trapped between the two surfaces. Having made this point, it must also be said that whereas many of the causative factors can be established with confidence as a result of careful questioning, dietary analysis and examination, others such as alcoholism, anorexia and bulimia may be confirmed only after a considerable amount of diplomatic persistence. Communication with the patient's medical practitioner is of particular importance in this respect.

Whatever the aetiology, one is always faced with the decision on whether or not to provide restorative treatment. Perhaps the most important consideration in this respect is the degree of worry expressed by the patient. Whereas extreme concern over a deteriorating appearance, painful or hypersensitive teeth or muscle dysfunction is likely to lead to early and active treatment, a wish to prevent further wear may well prompt a more circumspect approach. It is important to establish the amount of wear, whether or not it is consistent with the age of the patient and whether it is continuing at a significant rate. Examination of study casts taken at different times can be extremely helpful in this respect.

Unfortunately, the teeth may be so severely worn that they cannot be rebuilt successfully by means of crowns. Often the only practical solution to the problem is the provision of an overdenture. It may also be a relatively simple solution because the amount of wear may mean that only minimal tooth preparation is necessary and that endodontic treatment is not required. Such circumstances will be found when the wear has been so gradual that there has been time for the deposition of a substantial layer of secondary dentine so that pulp exposure has not taken place.

The potentially unfavourable complete denture
Many more patients experience problems with complete lower than with complete upper dentures and longitudinal studies have shown that the rate and degree of bone loss following extraction of the teeth is substantially greater in the mandible. The longest surviving units of the natural dentition are frequently the lower anterior teeth. If some of these teeth can be used as abutments, the subsequent overdenture has a valuable part to play in preventing the most unfavourable post-extraction changes in the mandible (fig. 2.7). Even if the abutment teeth are eventually lost, it can still be argued that the overdenture has assisted the

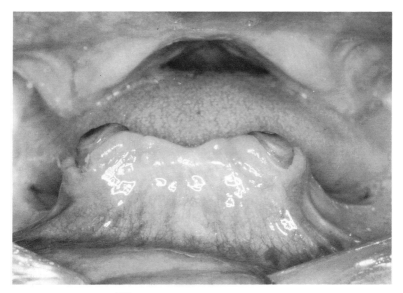

Fig. 2.7 Retention of 3|3 as root abutments has helped to maintain alveolar bone in the anterior region of the mandible. There is evidence of gross resorption in all other areas of the mouth.

patient to make the transition to the completely artificial dentition.

There is evidence of a decline in the rate of tooth loss in the adult population and, as a consequence, many patients face the transition to the edentulous state rather late in life, and often without previous successful experience in the wearing of partial dentures. The contribution which root abutments can make to the support and stabilisation of complete dentures offers a considerable benefit to these patients.

The doubtful partial denture abutment
There are occasions in planning restorative treatment, which includes the provision of partial dentures, when teeth which occupy a strategic position in the dental arch are found to be unsuitable for use as conventional abutments. This may happen if the teeth are over-erupted, if they have been severely damaged in their physical structure, or if they have lost so much periodontal support that they could not withstand the forces to which they would be subjected in helping to support and retain the dentures (fig. 2.8).

Retaining these teeth as overdenture abutments may avoid unilateral or bilateral free-end saddles or may provide valuable tooth support for what would otherwise be a very long bounded saddle.

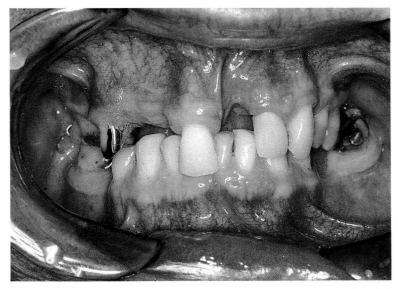

Fig. 2.8 The combined effects of alveolar resorption and over eruption have left 1| in a vulnerable position. It is no longer suitable for use as a partial denture abutment but could contribute as an overdenture abutment to the support of the long anterior saddle.

Patient selection
Overdenture treatment offers a potential benefit to all patients whose clinical requirements fall within the categories previously outlined. It is important to recognise, however, that wearing an overdenture places certain demands on the patient with respect to the maintenance of the dentures and, more importantly, of the abutment teeth, gingival tissue and adjacent mucosa. If the standard of oral hygiene is inadequate, it is likely that the abutments will fail and for this reason particular care must be exercised in the selection of patients for treatment.

Age
Young patients facing tooth loss due to uncontrolled caries or periodontal disease would clearly benefit if a number of teeth could be retained for use as overdenture abutments. However, such patients are in the high-risk category and unless the deterioration can be brought under control any benefit is likely to be of short duration. On the other hand, young patients who require a prosthesis for the restoration of a congenital or acquired defect in jaw structure, and for whom it is critical that tooth loss be prevented, are often ideal candidates for overdentures (fig.

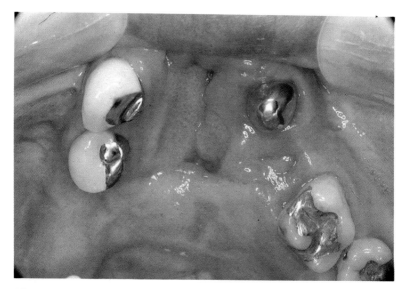

Fig. 2.9 This young patient has a repaired cleft of lip and hard palate. 43| have been restored with bonded porcelain crowns containing rest seats for a cobalt-chromium partial denture. The incompletely erupted |3 has been covered with a gold coping so that it can be used as an overdenture abutment.

2.9). It should be recognised that the teeth will be subjected to coverage by a denture for many years and they should therefore be protected by means of gold copings.

At the other end of the age scale, provided that the patient is fit and alert and able to cooperate in the necessary oral hygiene procedures, overdenture treatment should be available when clinically indicated.

State of general health

The provision of overdentures is inappropriate for patients with severe debilitating conditions where medical treatment is likely to be prolonged and patient cooperation uncertain. Mental handicap which prevents the patient from maintaining an adequate standard of oral hygiene would also be a contra-indication to this form of treatment. Physical handicap *per se* is not a contra-indication provided that plaque control can be achieved, and an overdenture may offer the best prospect of a stable prosthesis for those who depend on oral control for certain manipulative skills.

As endodontic therapy is a prerequisite for overdenture treatment in the majority of cases, any medical condition which contra-indicates root canal treatment might eliminate the possibility of overdentures. Patients who are candidates for cardiac surgery, for example, must be assessed very carefully and, unless an outstandingly good prognosis can be established for the proposed restorative treatment, it is probably best to avoid overdentures.

History of previous dental treatment

It is important to obtain as complete a picture as possible of the previous pattern of attendance for dental treatment and of the history of such treatment. An assessment should be made both of the quality of previous treatment and of the patient's standard of oral hygiene. Repeated failure by others to achieve successful long-lasting restorations should always sound a warning, and while it may not rule out a treatment plan involving overdentures, it may be an indication for treatment to be approached more cautiously. For example, an overdenture retained by precision attachments should not be attempted before ample time has been allowed for the assessment of such factors as the success of endodontic or periodontal procedures. Where excessive loss of face height is to be restored, it will be necessary to gauge the adaptation to an increase in occlusal height while a transitional prosthesis is in service.

Attitude to dental treatment

It is also very important to assess the patient's attitude to, and expectations of, dental treatment. Unless the patient is willing to accept the part that he or she has to play in achieving a successful outcome, it is very likely that the treatment will fail. Certain preliminary treatment, which may take the form of endodontics, periodontal surgery or crown preparation, necessarily precedes overdenture construction. This form of prosthesis therefore entails a number of extra visits and added expense.

Once the treatment has been completed the patient will be required to achieve a very high standard of oral hygiene and to maintain this standard throughout the life of the restoration. Thus, there must be a requirement for regular review and for periodic modification and eventual replacement of the prosthesis.

It is essential that patients should be made aware of these requirements at the outset. If they find the additional treatment proposals daunting or are unable to attend for all the necessary preliminary stages, then it may be wiser to consider a simpler alternative. If there is any doubt as to a patient's ability to attain the necessary standard of plaque control, a course of oral hygiene instruction should be instigated and performance monitored over a period of time before overdenture treatment is commenced.

3

Abutment Selection: Sequence of Treatment

In Chapter 2, the indications for overdentures and the criteria for patient selection were discussed. In this chapter the question of abutment selection is considered and the sequence of treatment for overdenture construction outlined.

Abutment selection

An abutment for an overdenture may consist of the modified crown or the modified root structure of a natural tooth. Where the crown structure is to be retained, it will be necessary for a substantial space to be created within the impression surface of the denture. It is unlikely that it will be possible to achieve this without either disrupting the occlusal plane or seriously weakening the denture unless there is sufficient space, arising from a considerable disparity in jaw size or relationship. As mentioned in Chapter 2, it is in precisely these situations, where anatomical factors are unfavourable, that this form of abutment is particularly valuable.

When most of the crown is to be retained, the tooth should be reduced to a tapering core shape and should be covered by a thimble-shaped gold coping in view of the high risk of caries.

In most instances root abutments can be formed by reducing the remaining crown structure to a rounded dome shape with a flattened central portion located approximately 2 mm above the level of the gingival tissues. The gingival margins of the preparation should be rounded and should be located at, or just above, the level of the gingival crest.

The primary function of the abutment is to contribute to the support of the denture. If additional retention is also desired, the preparation of the abutment may be modified in order to accept some form of attachment as described in Chapter 5.

21

Table 3.1 Percentage of lower teeth which are either sound and untreated or filled

Tooth type	Age 35-44	45-54	55+
3|3	98	96	88
4|4	91	84	70
5|5	76	67	47
7|7	70	56	33

Table 3.2 Percentage of upper teeth which are either sound and untreated or filled

Tooth type	Age 35-44	45-54	55+
1|1	87	77	61
3|3	90	81	61
4|4	75	63	41
5|5	75	59	36
7|7	75	61	38

The choice of abutments is, of course, dictated by the number of teeth remaining and by the position they occupy in the dental arches. In this respect it is worth looking at the prognosis of certain groups of teeth which are likely to remain, whether sound and untreated or satisfactorily restored. The evidence presented in Tables 3.1 and 3.2 is obtained from the adult dental health survey conducted in the UK in 1988. Lower incisors have been excluded as they are more difficult to root treat and have a limited capacity for load-bearing. The figures show conclusively that the lower canines and first premolars are most likely to be available as potential abutments as age advances. The prognosis of the corresponding upper teeth is generally worse although canines and central incisors have a good survival rate. These findings must be reassuring as there is theoretically a greater likelihood of lower teeth being available to guard against resorption of the underlying bone in the more vulnerable jaw. There are, however, a number of factors which must be taken into account in assessing the suitability of any tooth for use as an overdenture abutment: periodontal status; number and location in the dental arch; integrity of tooth structure; suitability for endodontic procedures; presence of associated bony undercuts; ability to offer positive retention; economics.

Periodontal status

The periodontal condition is a critical factor in the assessment of potential overdenture abutments. Where it is intended to retain the crown structure of the abutment tooth, it must be recognised that the tooth will still be subjected to considerable lateral and antero-posterior loading and is unlikely to survive unless it has substantial alveolar bone support and a healthy periodontium. Where teeth are to be cut down and used as potential root abutments these stringent criteria can be relaxed to some extent. Although mobile teeth with advanced bone loss are unsuitable, teeth with minor degrees of mobility can be used as root abutments, as reduction of the clinical crown will produce a more favourable crown:root ratio. Loading from the denture base is also more favourable, as elimination of coronal structure reduces the more damaging lateral forces. It is difficult to lay down rules on the least amount of investing bone which should remain before a tooth can be considered as a suitable abutment. As a guide, it is reasonable to say that if half the root is surrounded by bone and pocketing has been eliminated successfully the root may be considered as a root abutment.

Active gingival or periodontal disease should be treated and eliminated before abutment preparation is attempted. If an overdenture is fitted over inflamed gingival tissue the situation will continue to deteriorate and failure of the abutment will eventually occur. Pocketing of more than 3 mm depth should be eliminated before abutment preparation is undertaken. It has been thought that if the soft tissues are to withstand the forces transmitted from the denture base, it is necessary to ensure that an adequate band of attached gingivae is present around each root abutment. If a tooth which would otherwise be a satisfactory abutment lacks sufficient attached gingivae, it has been recommended that periodontal surgery, such as a free gingival graft or apically repositioned flap, should be performed. However, the evidence for such a recommendation is not strong and there is a view that favourable tissue response to overdenture wearing is dependant less on the actual gingival architecture and more on measures which eliminate gingival inflammation and ensure meticulous plaque control.

Number and location in the dental arch

Where possible, in the construction of a complete overdenture, bilateral abutments symmetrically located in the dental arch should be used. Such an arrangement will offer the most satisfactory support for the denture and will minimise the possibility of fatigue fracture of the denture base or of excessive loading of any particular abutment. If possible, a gap of at least one tooth width should be left between abutments to simplify oral hygiene procedures and avoid trauma to the interdental tissues. An exception to this general rule may be made when the denture is being provided to compensate for excessive loss of tooth structure. It may be wiser to retain a contiguous group of teeth to support the denture, especially if root canal treatment is not required in the preparation of the abutments (see fig. 3.5).

Canines are frequently among the last teeth to be lost and their large root surface area makes them particularly suitable for use as overdenture abutments. The bulbous form of the adjacent alveolar process may occasionally interfere with the insertion of the denture base, but it is usually possible to effect a compromise by reducing the extent of the flange without seriously affecting the retention or stability of the overdenture. Molars also offer substantial support, but their use as root abutments is limited by the fact that endodontic procedures are more complex and time-consuming. Occasionally, however, it may be possible to section a molar and to retain a single root as an overdenture abutment when either endodontic or periodontal considerations would render the whole tooth unsuitable.

Central incisors will generally be preferred to lateral incisors in the upper jaw. Lower incisors are probably the least acceptable overdenture abutments, although in the absence of a lower canine the preservation of an incisor may in some instances be the only way of avoiding a dental clearance. Premolars are generally satisfactory abutments, though if root canal treatment is necessary second premolars are preferred as they more commonly have a single root canal. They also offer good load distribution and simplify oral hygiene maintenance if canine abutments are also present.

Integrity of tooth structure

If a tooth is to be considered for use as an abutment retaining the bulk of the crown structure, it is essential that it be either intact or soundly restored. All carious lesions should be eliminated and any suspect restorations should be replaced. As mentioned earlier, the ideal form of a root abutment is a dome shape with margins established just supragingivally (fig. 3.1). If it is possible to complete this preparation with smooth regular margins and an intact surface to the dome, it should be possible for the patient to maintain the abutment without the need for coverage by a coping. This will be the case with the vast majority of root abutment preparations. Where it is not possible to establish a smooth shape and regular margins, either because of caries or previous

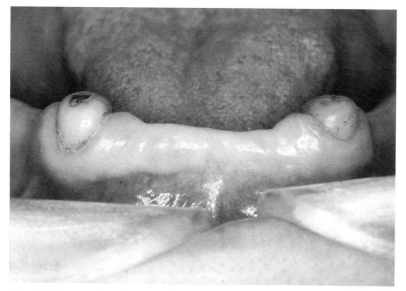

Fig. 3.1 Dome-shaped preparations on root-treated mandibular canines.

restorations, it will be necessary to cover the root face with a gold coping (figs 3.2 and 3.3). It may sometimes be necessary to undertake gingival surgery to place the margins at a level where the patient can reasonably be expected to undertake the necessary oral hygiene procedures. Teeth with root caries, or with surface damage which is not amenable to these restorative procedures, are unsuitable for use as root abutments.

Suitability for endodontic procedures
Most potential root abutments will require endodontic treatment before preparation of the root face can be undertaken. It is therefore necessary to determine whether there are any factors which may complicate, or indeed prevent, successful root canal treatment (fig. 3.4). Accurate periapical radiographs are essential for assessment of root canal morphology.

In the age group of patients for whom overdentures are most commonly constructed, the combined effects of caries, previous tooth preparation and wear of tooth substance, together with the normal ageing reaction of the dental tissues, may have resulted in considerable reduction in the volume of the coronal and radicular pulp structures.

If sufficient secondary dentine has been laid down to obliterate the pulp chamber, and if preparation of a domed root face can be undertaken

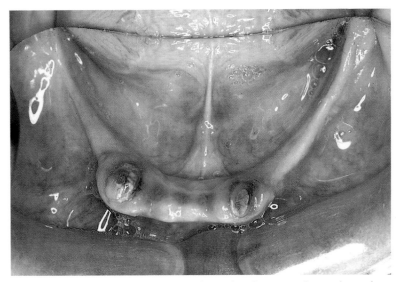

Fig. 3.2 Margins of abutments are irregular due to caries and previous restorations and coverage will be required.

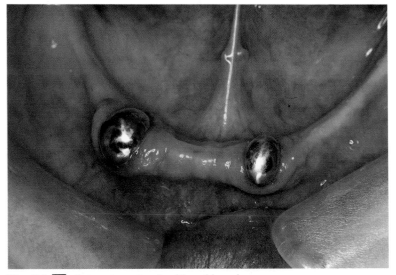

Fig. 3.3 3̄1̄2̄ abutments restored with gold copings.

without causing sensitivity or a pulpal exposure, then there is no need to undertake root canal treatment (fig. 3.5). Such teeth should be kept under regular review, however, and endodontic treatment will be

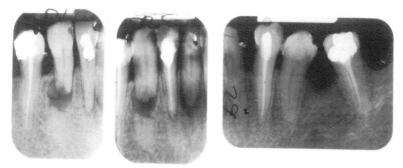

Fig. 3.4 3̅| is unsuitable for endodontic treatment due to root fracture and resorption. 4̅2̅|3̅5̅ have been root-treated for use as root abutments to provide wide distribution of support for the overdenture.

necessary if pulpal symptoms arise. Serious consideration should be given to removing a vital pulp if its presence compromises correct tooth preparation. Constrictions in the apical region may restrict instrumentation and may prevent successful endodontic treatment. However, much of the endodontic treatment associated with overdenture abutments involves elective devitalisation of previously symptom-free teeth.

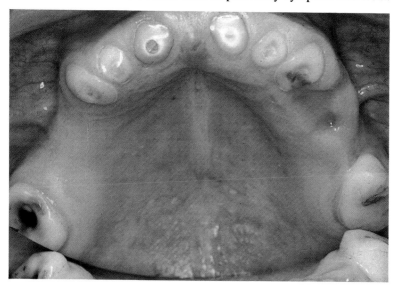

Fig. 3.5 Severely worn 3̲2̲1̲|̲1̲2̲3̲ have been prepared as root abutments. Only 1̲| required root canal treatment. |̲3̲ has an area of stained dentine and is concave in contour. Restoration by means of a gold coping will probably be required at a later date.

In these situations, if the pulp contents can be successfully removed and the canals prepared and filled to a consistent working length, even if this is just short of the true length of the root, successful root treatment will generally be achieved. It is necessary, of course, to monitor the apical condition by regular post-operative radiographs.

It has already been mentioned that canines are probably the most frequently used root abutments, and fortunately they usually present few obstacles to the successful management of root canal treatment. Where a choice of abutments exists, preference from an endodontic viewpoint would be given to those with a clearly defined root canal, with straight or gently curved roots and with single rather than multiple canals and to those without a pre-existing apical lesion.

Before concluding this section on endodontics it is worth reporting on success rates. In a recent study of 679 abutment teeth a failure rate of only 3.8% was recorded over a 12-year period. Of particular significance is the fact that over half the failures were the result of pulpal necrosis of teeth which had not received endodontic treatment. It was concluded that preparing the root faces of vital teeth may create microexposures of the pulp remnants. As the risk of pulpal necrosis to these teeth was shown to be 1 in 5, it is recommended that the remains of the root canal, even if apparently calcified, should be sealed with a restoration such as glass ionomer cement.

In the same study it was noted that nearly a quarter of the abutment teeth lost their coronal restoration, and consequently the coronal seal, because of carious attack. Nearly 1 in 10 of the failures was due to a 'perio-endo lesion' which resulted from poor oral hygiene. As in so many instances, success of this element of treatment is heavily dependant upon effective oral hygiene together with regular recall and review.

Presence of associated bony undercuts

It is important to record impressions and obtain study casts of the dentition before finalising the treatment plan and abutment selection. The study models should be surveyed to establish the location and the extent of undercut on the natural teeth, on the edentulous alveolar ridges and particularly on those areas of the alveolar process around potential overdenture abutments.

The labial contours of the alveolar process may sometimes be so prominent, particularly in the canine region, that a marked undercut area is present in the buccal sulcus. This bony prominence may often prevent the insertion of a denture with a complete labial flange (fig. 3.6). It will usually be possible to reduce the extent of the denture base

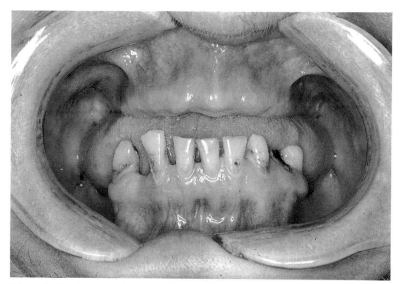

Fig. 3.6 Careful analysis of the bilateral bony undercuts in the mandible will be necessary prior to finalising the choice of abutments and design of overdenture.

without seriously prejudicing the stability of the overdenture. In some instances, the alveolar bone may be so prominent that the facial contours of the patient would be distorted by fitting even a reduced labial flange. It is therefore occasionally necessary to reject the potential abutment teeth where it is judged that their removal and subsequent remodelling of the alveolar process would produce a situation more favourable for the construction of a stable and aesthetically satisfactory prosthesis.

Ability to offer positive retention

By retaining root abutments it is possible to retain alveolar bone and produce dentures which are stable and well supported. That this has also contributed to the retention of the dentures is seen as an additional bonus even if it was not a primary objective of the technique.

On some occasions, however, there is a requirement for additional mechanical retention, particularly when anatomical features are unfavourable or when the patient has a strong psychological need for positive retention of the prosthesis at all times. Where attachments are envisaged, it must be recognised that the abutment tooth will be subjected to considerable forces during insertion and removal of the overdenture as the retentive device engages and disengages; it will also be subjected to increased lateral and antero-posterior forces during normal function.

Comment has already been made on the need for the most stringent evaluation of the periodontal status of abutments whose crown structure is to be maintained. These comments apply even more forcefully if the teeth are to be used to carry some form of precision attachment. Many teeth which would serve quite successfully as root abutments for overdentures would not be suitable for this additional task and might fail if subjected to these additional forces.

Where suitability is questioned, it is probably wise to delay the decision on attachments until the original overdenture has been in service for at least 6 months. This time lapse allows an assessment of the response of the tooth and adjacent tissues to loading under the denture base, and for any necessary modification to the denture or to the abutment preparation. The period should be increased if the fitting of the overdenture has been combined with the immediate replacement of any of the adjacent teeth.

Economics

It must, of course, be recognised that the selection of numerous abutment teeth may entail additional endodontic treatment or the provision of additional gold copings. The patient will certainly be required to attend for more treatment sessions. All this adds, inevitably, to the cost of treatment and it will often be necessary to compromise on economic grounds between an ideal and an acceptable number of abutments for the construction of a satisfactory overdenture.

Sequence of treatment
Examination, diagnosis and treatment plan

The treatment plan should be based on a thorough clinical examination supplemented by a careful history of relevant medical, dental and personal factors, as discussed in Chapter 2. The structural integrity of the proposed abutment teeth should be ascertained and their response to vitality testing noted. Pocket depth and mobility should be measured and the extent of attached gingivae noted. Radiographs are essential for the assessment of bone support and for endodontic evaluation. Study casts should be obtained and surveyed in order to assess the degree of undercut in relation to potential abutment teeth, extent of denture borders and path of insertion. Suitable abutments should be selected in line with the criteria previously described and an outline treatment plan should then be presented to the patient.

It is important that the patient understands fully what is to be done and appreciates that the crown of the tooth is to be removed. To some, this tooth preparation might be seen as unacceptable mutilation of the little that remains.

Periodontal treatment

Periodontal treatment should be initiated as soon as possible. It is important to assess the extent of disease and to set up a monitoring system which is readily understandable by both patient and dentist so that the effect of treatment can be gauged. Indices which measure presence of debris, pocket depth and bleeding on probing are particularly valuable in this respect.

It is not the purpose of this book to provide details of periodontal treatment as they can be found elsewhere. However, it is important to make the basic point that successful treatment involves efficient monitoring and enthusiastic motivation not only during the first course of treatment but for as long as the patient is the personal responsibility of the dentist.

Effective control of periodontal disease is more than likely going to require a chance in the patient's behaviour. The following model has been described for accomplishing this change.

(a) Find out what the patient wants to achieve in the long term regarding dental and oral health.

(b) Provide the relevant information which will enable the patient to understand the disease process and how the proposed treatment will improve the situation.

(c) Monitor the patient's attempts to achieve these personal objectives. The use of indices allows the patient to play an active part in an atmosphere which is not seen as being threatening or authoritarian.

(d) Where necessary, help the patient to achieve the change in behaviour in order to effect further improvement.

(e) Continue the monitoring and motivation procedure over the long term.

Preliminary treatment of abutments

Root canal treatment should be completed early in the sequence of treatment so that there is a reasonable opportunity to evaluate the success of endodontic procedures before embarking on final preparation of the root abutments. Any periodontal surgery which is necessary in the area of the abutments should also be completed at this stage.

Preparation of the abutment teeth

Teeth whose crown structure is to be retained should be tapered in all dimensions to allow for the fitting of a thimble-shaped gold coping.

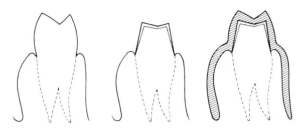

Fig. 3.7 Diagram of the stages in preparation of a thimble-shaped gold coping and its relationship to the overdenture.

They should also be reduced occlusally to provide for an adequate thickness of overlying denture base material (fig. 3.7).

Root abutments should take the form of a smooth, rounded dome preparation with a flattened central portion located 2 mm above gingival level. When this can be achieved without resorting to root canal treatment, it is quite sufficient to polish the prepared abutment with prophylactic paste, thus leaving the completed preparation easily accessible for hygiene procedures. When root canal treatment has been necessary the only modification to the dome preparation is the addition of a restoration to seal the access to the root canal. Amalgam or glass-ionomer cement are suitable materials for this purpose.

If a gold coping is required, it should reproduce the requisite dome

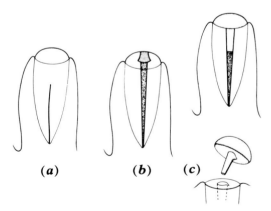

(a)   *(b)*   *(c)*

Fig. 3.8 Diagram of different forms of dome preparation. (a) Dome-shaped preparation without endodontic treatment where the pulp chamber and root canal have been almost obliterated by secondary dentine. (b) Root-treated abutment; root face restored with amalgam or glass-ionomer cement. (c) Gold coping to establish correct dome-shaped contour and incorporating short post and antirotational notch.

shape and should, if possible, have its margins at, or just above, gingival level. If the tooth has been root-treated, the coping should incorporate an extension into the root canal to a depth of approximately 3 mm for additional retention, together with an antirotational notch (fig. 3.8).

Denture construction

When all the necessary preliminary treatment procedures have been completed, the normal stages of denture construction are followed through and these will be considered in more detail in Chapter 4. The precise timing of the final abutment preparation in relation to prosthetic procedures will depend on whether or not it has been decided to construct an immediate overdenture.

Follow-up and maintenance

After the overdenture has been fitted, regular review and maintenance are required. The patient must maintain a very high standard of oral hygiene if the health of the abutment teeth and gingival tissues is to be assured and the clinician must supervise this aspect of performance and must also be ready to undertake modifications to the dentures and abutments as these become necessary. The subject of maintenance is dealt with in detail in chapter 6.

4
Clinical and Laboratory Techniques

This chapter is restricted to a consideration of the dome-shaped preparation which is the simplest and, perhaps, most popular of abutment preparations. The relevant clinical and laboratory techniques involved in the production of immediate, conversion and replacement overdentures are presented.

Immediate complete overdentures
The obvious attractions of the conventional immediate denture are equally applicable to the overdenture. The transition from the natural to the completely artificial occlusal surface frequently involves the extraction of teeth and the preparation of overdenture abutments at the stage when the dentures are fitted. The following description deals with the clinical situation in which the remaining natural teeth are $\overline{321|123}$ and a decision has been made to retain $\overline{3|3}$ as the overdenture abutments (fig. 4.1). Cast metal copings would not normally be contemplated for the immediate restoration.

Preliminary clinical stages
Following appropriate periodontal and endodontic treatment, the stages of impression taking and recording the occlusion are completed as routine procedures. As a labial flange will be provided wherever possible, it is particularly important to obtain an accurate impression of the labial sulcus. The jaw relationship in the horizontal plane should be recorded with the mandible in the retruded position, a position not influenced by the sensory feedback of an already mutilated natural dentition.

At this stage the point should perhaps be made that all the principles of complete lower denture design should be fully implemented and that the additional support and stability afforded by the abutment teeth

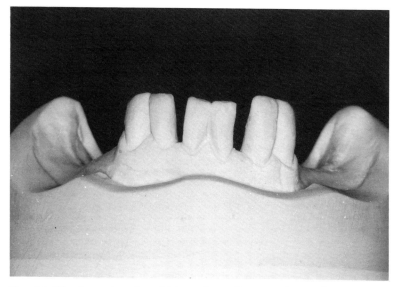

Fig. 4.1 The lower jaw for which an immediate overdenture is to be constructed.

should be regarded as a welcome bonus. If the denture base does not cover the maximum area, if the occlusion is not carefully balanced, or if the denture is not placed in the neutral zone, movement of the denture will irritate the gingival tissues. The unnecessary increased loading of the abutment teeth may well encourage destruction of the supporting tissues. Thus the benefits of the overdenture are put in jeopardy.

Of course, load distribution in the upper jaw is rather less critical and it may be possible to reduce palatal coverage if overdenture abutments are retained. Indeed such an approach to denture design may be adopted when treating the patient who has a tendency to retch and who may only tolerate a smaller connector. Although the presence of overdenture abutments will improve stability, it may be necessary to augment retention to compensate for the deficiency in base extension and border seal, by using attachments (see Chapter 5).

Trial dentures

At the try-in stage, in addition to a routine examination of the trial dentures, an assessment is made of the amount of bone which surrounds the teeth to be extracted; evidence from radiographs and from measurements of pocket depth will assist in the correct preparation of the cast. It is not the purpose of this book to discuss immediate denture technique *per se*, but two points should be made. First, it is often possible to provide a flange without modifying the contours of the residual ridge, especially in the lower jaw; even in the upper jaw, when the alveolar bone has a pronounced labial undercut, it may still be possible to incorporate a partial flange. Second, space for a flange has often been created by the bone loss of periodontal disease; once the teeth have been extracted rapid collapse of the sockets will provide all the room that is necessary. This collapse should be anticipated by cast trimming.

The position of the natural teeth within the arch, and in relation to the opposing dentition, should be evaluated. As a consequence of periodontal disease, the teeth may be splayed out and over-eruption may have occurred. The position of the replacement artificial teeth may need to be modified and details of the proposed changes should be noted so that the technician has an accurate prescription from which to work. In an ideal world, it is probably best that the clinician discusses the changes directly with the dental technician. In the absence of a dental laboratory or technician on the surgery premises, the clinician should decide whether or not it is wiser to trim the cast and set up the anterior teeth himself.

Cast preparation

Before preparing the overdenture abutments, or removing the teeth that are to be extracted, it is helpful to produce a labial index which will serve as a permanent record of the existing tooth position and ridge shape. This index can be conveniently made from silicone putty (fig. 4.2).

The overdenture abutments are carved down to a level which will be slightly higher than the eventual preparation in the mouth. The distance from gingival margin to apex of the model preparation is measured and recorded in the patient's notes (fig. 4.3). This measurement will be a useful guide when mouth preparation is carried out at the next visit. Acrylic teeth of appropriate size can then be fitted by hollowing out the under-surface and fitting over the preparation (fig. 4.4).

The incisors should now be removed from the cast one at a time and the artificial teeth waxed in position. It is at this stage that any change in position is made. The putty index enables the magnitude of change to be monitored carefully (fig. 4.5).

Once the acrylic teeth have been positioned and secured to the trial denture with wax, the residual ridge in the $\overline{21|12}$ region should be modified to mimic the change expected following extraction. In the lower jaw it is wise to avoid surgical removal of bone unless, perhaps, a gross labial undercut has to be eliminated. The lower ridge has the unfortunate habit of resorbing only too quickly without any help from

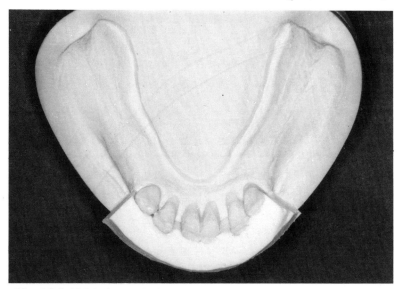

Fig. 4.2 Labial index made from silicone putty.

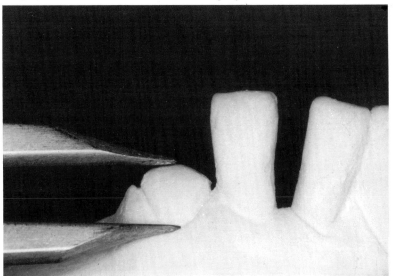

Fig. 4.3 Abutment shape prepared on cast. Height of preparation measured with dividers.

a bur or rongeurs. Thus the cast trimming may well be restricted to smoothing the crest of the ridge and removing cast material on the labial aspect to a depth of 0.5–1 mm (fig. 4.6). Again, the labial index acts as

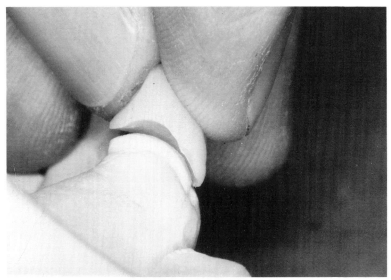

Fig. 4.4 Artificial tooth fitted over the prepared abutment.

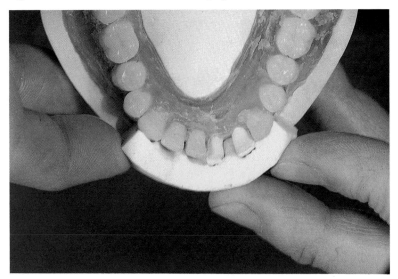

Fig. 4.5 $\overline{21}$ replaced by artificial teeth which have been positioned further lingually.

a useful guide as the dental stone is carved away. Once the occlusal positions of $\overline{321|123}$ have been checked, the labial flange can be added in wax. The immediate overdenture is now ready for processing.

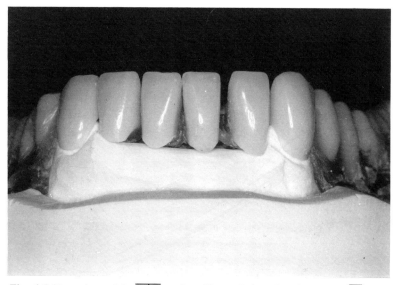

Fig. 4.6 Cast trimmed in $\overline{21|12}$ region. Note relationship of artificial $\overline{3|3}$ over prepared abutments.

As more tissue will remain in the region of the overdenture abutments, the denture will be consequently thinner in cross-section. If findings from the original assessment of the patient suggest that fracture of the denture base is a great risk, it may be wise to process the denture in a high-impact resin.

Problems may be encountered when complete overdentures have abutments only in the anterior region. Because the anterior portion is supported firmly on tooth substance and the posterior portion on more compressible mucosa, the whole appliance will not 'bed' into the mucosa evenly, with the result that the posterior border seal may be less effective. To overcome such a problem in the upper jaw, a wider post dam may be carved into the model so as to improve the contact zone. An alternative approach, in the lower jaw, is to use the special impression technique described in Chapter 5.

Fitting the overdenture

The first stage of treatment is to prepare the overdenture abutment teeth. The gross removal of the crown may be achieved with a long diamond fissure bur, the horizontal cut being made at approximately the height to which the tooth on the model was prepared. For security, a hole may be cut through the body of the crown close to the tip; a piece of dental

Fig. 4.7 A technique for removal of the crown of an abutment tooth.

floss is looped through the hole, allowing the dental surgery assistant to hold the piece of tooth as it is severed from the root (fig. 4.7). Following this gross removal, the root face is ground to produce the domed shape at a level slightly less than that of the preparation on the cast to which the denture was processed. The measurements made at the model preparation stage act as a useful guide in this respect. Once the tooth surface has been smoothed, the entrance to the root canal is sealed either with amalgam or a glass-ionomer cement. Having prepared the canine teeth, the incisors are extracted and any necessary ridge adjustment carried out. The overdenture is then fitted and the patient discharged. It will be appreciated that at this stage the denture is not fitting accurately on the root faces.

Immediate recall procedures

Ideally, the patient should be seen the day after the extractions. The restoration will now be sufficiently hard for the whole tooth surface to be polished and the overdenture fitted accurately. Accurate adaptation of the denture may be achieved with cold-curing acrylic resin as follows.

Using a small round bur, a vent hole is cut from the lingual aspect of the polished surface of the denture through to the impression surface. The oral mucosa may be protected by a thin film of petroleum jelly before cold-curing resin is poured into the impression surface; the denture is replaced in the mouth and retained in position by light occlusal pressure. Any excess

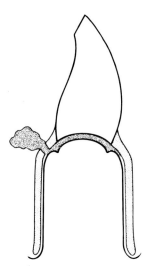

Fig. 4.8 Diagrammatic representation of technique used to adapt overdenture to the abutment tooth. Excess resin is shown escaping from the vent.

will escape through the vent (fig. 4.8). Once the resin has reached the dough stage, the denture can be removed and the curing cycle allowed to continue outside the mouth. When the resin has hardened, any thin flash which has been pushed into the gingival crevice should be removed. The denture can then be fitted and the occlusion re-checked. Further maintenance procedures are discussed in chapter 6.

In concluding this section the point should be made that the immediate overdenture technique employs straightforward clinical and technical procedures. The finished restoration exhibits all the advantages of the conventional immediate denture. As the denture receives positive support from the abutment teeth, there is little likelihood of it sinking into the healing tissues, with the consequence that a reasonable level of comfort should be guaranteed.

Conversion overdentures
Occasionally it is desirable to convert an existing partial denture to an overdenture rather than to construct a totally new appliance. The conversion can be undertaken by either of the two following techniques, depending upon whether there are laboratory facilities at the surgery.

Chairside conversion
In figure 4.9 the lower partial denture replaces all teeth except the canines. The retention of these teeth is valuable and their prognosis is

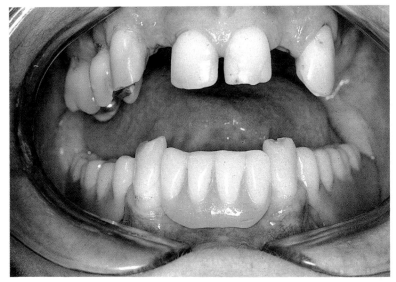

Fig. 4.9 Lower partial denture prior to chairside conversion.

considered to be more favourable if incorporated in the overdenture rather than being subjected to the stresses of a tissue-borne partial denture. The existing denture is being worn with satisfaction.

The technique to be described possesses the advantage of convenience and economy. However, as the additions to the denture will be made in cold-curing resin, it should be appreciated that the weaker structure might be prone to fracture.

When the endodontic treatment of the canines has been completed, an impression is taken of the arch with the denture *in situ*; the denture should be retained in the impression. If alginate is used, the impression should be kept in a moist atmosphere during subsequent procedures. Alternatively, silicone putty may be employed (fig. 4.10). The natural teeth are now prepared as overdenture abutments, the surfaces polished and restorations inserted into the entrances to the root canals. Cold-curing resin of appropriate tooth shade is poured into the tooth indentations of the impression, and the impression, including the denture, seated carefully in the mouth. Once the resin is sufficiently hard, the impression can be removed and polymerisation allowed to continue outside the mouth. The denture with added teeth can then be removed carefully from the impression and refitted in the mouth (fig. 4.11). A flange can be added conveniently with one of the cold-curing materials designed to be used in the mouth. In this respect, it is worth

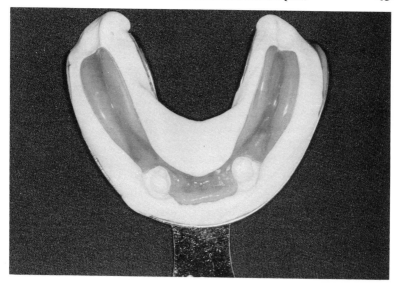

Fig. 4.10 Lower partial denture incorporated in the silicone putty impression.

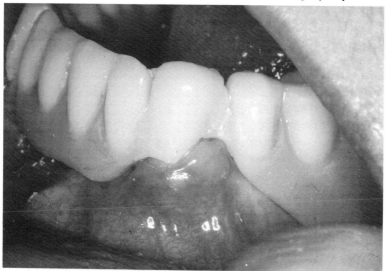

Fig. 4.11 Conversion overdenture: addition of $\overline{3}$ to the existing partial denture prior to adding flange and final occlusal adjustment.

noting that certain poly(butyl methacrylate) chairside materials do not cause mucosal irritation, nor do they possess as high an exotherm as some of the poly(methyl methacrylates). Once the flange has hardened,

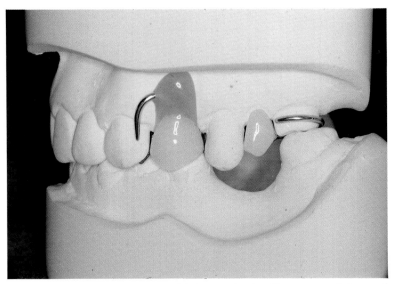

Fig. 4.12 Casts showing upper partial denture *in situ*. The over-erupted second premolar is to be converted into an overdenture abutment.

the occlusal contact of the added teeth should be checked and the additions polished.

Using the above technique the complete conversion has been undertaken in the surgery with minimal inconvenience to the patient.

Laboratory conversion
The following technique may be employed where laboratory facilities are immediately available. The example used in this instance is shown in figure 4.12. The single premolar interrupts what otherwise would be a long saddle for the existing cobalt-chromium partial denture. In the absence of denture support the tooth has over-erupted and there has been some loss of periodontal attachment. As there is still sufficient attachment for the tooth to be used as an overdenture abutment, the decision is made to root-fill the tooth and convert the denture so that the long bounded saddle can be provided with useful additional support in the middle of the span.

The premolar is prepared as a dome-shaped abutment, the tooth surface polished and the restoration placed in the entrance to the root canal. An impression is taken with the partial denture *in situ*; an impression of the opposing arch and a recording of the occlusion is also obtained.

Once the casts have been poured, an artificial tooth is fitted to the replica of the root face and placed in a more satisfactory occlusal position. A flange is added and the denture refitted in the patient's mouth.

Replacement overdentures

Occlusal wear is perhaps the most common reason for replacing overdentures. This is not altogether surprising when we recall that the concept of overdenture treatment affords a higher degree of masticatory efficiency. Perversely, the same point may be partly responsible for another reason for replacement, namely fracture of the denture base. Rapid wear of the occlusal surfaces results in loss of occlusal vertical dimension, which leads to a change in the relationship of the anterior teeth with the potential for increasing the load transmitted to the anterior part of the denture base. Wear of the posterior teeth may result in the well-recognised wedging action which adds to the level of stress in the midline of the palate. In circumstances where fracture of the denture due to rapid occlusal wear has become a particular problem, metal occlusal surfaces can be constructed so as to stabilise the occlusal relationship as effectively and for as long as possible. Fracture is particularly likely to occur when the overdenture is subjected to the high loads generated by opposing natural teeth — another important indication for overdenture treatment, as mentioned in Chapter 2.

On occasions it is necessary to reshape the abutment teeth because of caries or gingival recession; both periodontal treatment and the provision of copings may have to be undertaken. An existing immediate overdenture may cease to fit accurately because of resorption at the extraction site. In some cases a replacement denture will then have to be constructed although the point should be made that an accurate fit of the denture base can be re-established by relining or rebasing.

A replacement overdenture can be made by conventional procedures or by using a copy technique. It may be necessary to incorporate a metal base to give added strength if the previous denture has fractured.

Conventional denture construction

Little need be said under this heading other than to make the point that it is frequently necessary to use an elastic impression material to obtain an accurate impression of the undercut ridges around the abutment teeth. A more detailed discussion of the varying compressibility of the tissues, and how this variation is likely to influence impression technique, is presented in Chapter 5.

Copy dentures

In recent years, a great deal of interest has been shown in developing copy techniques, following the increasing realisation that successful adaptation to replacement dentures is partly dependent upon the faithful reproduction of a successful polished surface.

Whatever copying technique is used the principle remains the same, namely that the copy serves as a matrix on which controlled modifications are made. For example, the unmodified copies serve as the record rims on which changes of the occlusal relationship can be made. The trial dentures are then constructed and, finally, the impression surfaces are modified by taking wash impressions in the trial dentures. In the presence of undercuts it is wise to use a light-bodied silicone elastomer as the impression material.

Metal bases

The only advantage gained by including a metal base is an increase in the rigidity of the denture base. Against this advantage should be weighed the disadvantages of increased cost, subsequent difficulty in relining and occasional problems of appearance. This latter point should perhaps be elaborated upon.

The bulkier ridge created by overdenture abutments means that inevitably there will be less space for the overlying denture. Lack of space may be apparent not only on the crest of the ridge but also on the buccal aspect. A metal reinforcement must be of adequate thickness to do its job properly and, at the same time, must not impede tooth placement nor show through the gingival margin area of the denture. The upper denture is probably the more critical in this respect.

Design of the upper metal reinforcement

Descriptions have been given in the literature of a metal base whose coverage is restricted to the abutment teeth and the crest of the ridge in the labial segment (fig. 4.13). Although it can be argued that such coverage is placed in an area prone to fracture, it may have to be so thin, so as not to compromise appearance, that it adds relatively little to the overall strength of the appliance. When inter-ridge space is limited it is probably better to provide complete palatal coverage to prevent mid-line fracture of the denture.

In the region of the abutment teeth it is advisable to place the metal base directly on the tissues, taking care not to extend it too far over the ridge. Relief may be provided on the crest of the ridge in the buccal segments, thus allowing acrylic resin to contact the mucosa and be

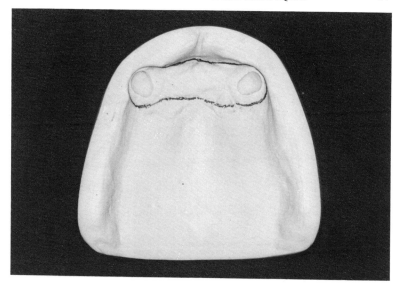

Fig. 4.13 Outline of metal reinforcement restricted to the labial segment. The disadvantages of this limited coverage are discussed in the text.

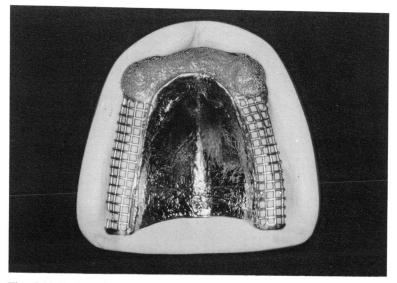

Fig. 4.14 Design of a full-coverage cobalt-chromium palate.

securely retained to the metallic base; the resin in the labial segment is retained by beads of metal (fig. 4.14).

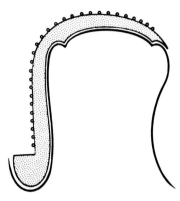

Fig. 4.15 Cross-sectional diagram of the design of a lower metal reinforcement. Note that the labial extent of the casting does not enter the undercut area.

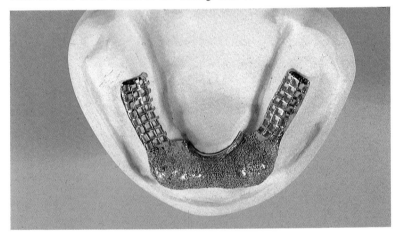

Fig. 4.16 Lower cobalt-chromium casting.

Design of the lower metal reinforcement

As the lower denture is prone to fracture in the mid-line, maximum reinforcement should be provided in the labial segment.

The metal casting should be in contact with the abutment teeth and the intervening ridge tissues. The metal should be extended into the lingual sulcus and the border constructed to occupy the full functional width and depth of the sulcus, so as to produce a more rigid girder structure (figs 4.15 and 4.16). Needless to say, a very accurate impression of the functional sulcus must be made. If labial undercuts exist, care should be taken to avoid the metal being extended into those undercuts.

5

The Use of Copings and Attachments

In Chapter 4, particular attention was paid to the simple dome-shaped preparation where the tooth surface is highly polished but otherwise unprotected.

In many cases this form of tooth preparation will function admirably in the long term and will thus be the preparation of choice not only for immediate overdentures but also for replacement overdentures. However, there are occasions where gold copings should be fitted or attachments provided. Cast gold copings may be required in the following circumstances.

1. Where caries of the root face of an existing overdenture abutment has developed.

2. Where the shape of the root face does not allow the orthodox dome-shaped preparation to be achieved.

Concavities in the tooth surface may arise from previous carious attack, cervical abrasion cavities and the result of gross attrition by the opposing teeth (fig. 5.1). Where there is no longer an adequate amount of tooth substance above the level of the gingival margin, very little stability will be afforded by the abutment teeth and consequently there will be relatively little protection of the gingival margins.

3. Where the abutment teeth are opposed by natural teeth in a patient who indulges in occlusal parafunction.

If the overdenture is not worn at night it is quite possible that the surfaces of the abutment teeth will be worn further by nocturnal parafunction or the roots may indeed fracture.

Attachments may be required in the following clinical situations.

1. Where additional mechanical retention is needed to overcome problems where the prognosis for denture retention is poor. For example, the retention of two overdenture abutments and the subsequent use of attachments may allow an adequately retained upper

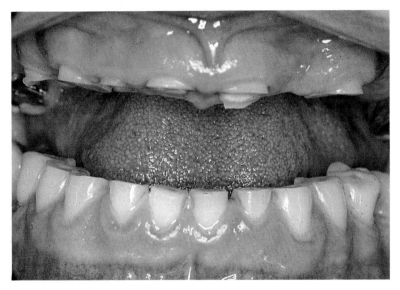

Fig. 5.1 Gross attrition of the upper anterior teeth in a 65-year-old man.

denture to be provided where there is little depth of sulcus, where the palatal shape is unfavourable or where the denture is opposed by natural teeth.

2. Where a labial flange cannot be provided and compensation has to be sought for the reduced retention.

3. Where the palatal coverage has been reduced in the treatment of retching problems. In this instance, the attachments will improve retention and also give the patient a greater feeling of security.

4. Where more positive retention is required for an extensive saddle on a partial denture (fig. 5.2).

5. Where it is desirable to splint together two abutment teeth by means of a bar attachment.

Copings

Both dome and thimble shapes can be used (fig. 5.3). The thimble-shaped preparation is, in essence, that which is used for a full veneer crown, with the margin being either a knife-edge or a chamfer. It will be appreciated that the walls of the preparation should have no more than a minimal taper so that adequate retention of the coping and the subsequent denture is assured.

The thimble shape offers more retention and stability for the overdenture than does the dome preparation and there is no requirement

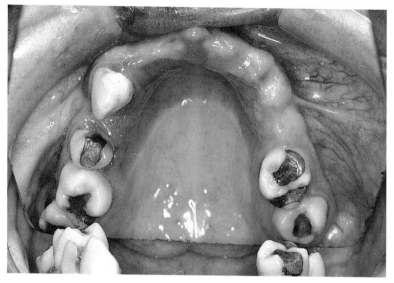

Fig. 5.2 3| will be prepared as an overdenture abutment to provide support. If necessary, an attachment may be added to provide additional retention for the extensive anterior saddle.

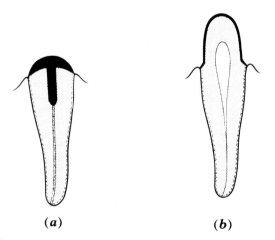

(a)　　　　　　　**(b)**

Fig. 5.3 (a) Dome-shaped coping; (b) thimble-shaped coping.

for endodontic treatment. However, protection of the tooth surface by a metal coping is mandatory and loading of the clinical crown is obviously greater. Further disadvantages arise from the longer clinical crown; it is likely to provide more of an obstacle to correct artificial tooth placement and the reduced bulk of the surrounding denture may lead to fracture of the denture base.

In contrast, the more favourable crown:root ratio of the dome-shaped abutment ensures more favourable loading conditions for the abutment tooth, a reduced risk of denture fracture and a greater freedom for correct tooth positioning. The simple shape of the dome preparation is likely to allow more efficient plaque control. As the tooth preparation is less conspicuous, a dome may provide less retention and stability. In addition, endodontic treatment will be required in the vast majority of cases. On considering the relative merits of the two preparations, the authors believe that the dome-shaped preparation offers more practical advantages as it is applicable in a larger number of clinical situations.

In normal circumstances it is inadvisable to fit a coping when an overdenture is first provided. Instead, if a few months are allowed to elapse it will be possible to gauge the patient's level of plaque control with greater confidence and the position of the gingival margin will have stabilised, allowing the margin of the coping preparation to be determined with greater certainty.

When preparing a tooth for a dome-shaped coping, sufficient tooth substance should be removed to allow for adequate thickness of the gold such that the apex of the dome will be no more than 2 mm above the level of the gingival margin. An antirotational notch (see fig. 3.8c) should be included and the margin of the preparation should be supragingival if possible. As the displacing forces on the coping will be minimal, it is not necessary to make provision for a long post. Impressions of the abutment teeth are obtained using conventional techniques.

Attachments

The particular value of attachments is the increase in retention afforded to the overdenture; clinical indications for such an approach have been listed earlier.

Attachments possess certain drawbacks when compared with the simple dome-shaped abutments (with or without copings).

1. The cost of treatment is increased.

2. Subsequent maintenance is likely to be more complicated.

3. The increased bulk of the attachment will weaken the overlying denture base and may predispose to fracture of the base.

4. Effective plaque control may be more difficult to achieve, especially with the bar attachments.

5. Higher loads are transmitted to the abutment teeth during masticatory function and also when the overdenture is inserted and removed.

Teeth selected to carry attachments must therefore have better periodontal support and be of such a shape that a substantial post can be
fitted.

As a consequence, many clinicians prefer to adopt the simpler line
of treatment and advance to attachments only if adequate retention has
not been achieved.

It is not the purpose of this book to consider attachments in any great
detail, as other texts serve this purpose admirably. Rather, the authors
are concerned with more general considerations.

Stud attachments

The many designs of stud attachment are based on the male/female
attachment principle with, in most cases, the male portion fixed to the
abutment tooth and the female portion embedded in the denture. The
attachments vary considerably in size and thus it is important, when
choosing the most appropriate system, to examine the available space on
articulated study casts. In this way the available inter-ridge space and the
correct inclination for the attachment can be assessed. If the attachments
are planned for complete overdentures, it is advisable to examine the
space available in the presence of correctly trimmed record rims so as to
judge whether a particular attachment is likely to prejudice correct
tooth positioning. Figure 5.4 shows a Ceka attachment within a sectioned denture. It can be appreciated how much space is required for this
particular attachment and that if it was inclined, either lingually or
labially, problems would arise in shaping the lingual contours of the
denture base or in aligning the artificial teeth correctly.

Most of the available attachments are designed to be fixed to gold
copings. Thus, when preparing the root face, sufficient tooth substance
must be removed to keep the overall height of coping and attachment
down to an acceptable level. In contrast to the tooth preparation for the
dome-shaped coping, more extensive preparation of the root canal
must be undertaken in order to achieve additional post anchorage.
Effective post retention is best obtained by using one of the prefabricated post systems which fit accurately into canals prepared by
matching reamers.

Examples of stud attachment systems

Typical of the ball and socket stud attachment is the Dalla Bona 604 (fig.
5.5), which allows for a certain amount of movement of the denture.
Such movement should be strictly limited by observing the correct
principles of denture construction.

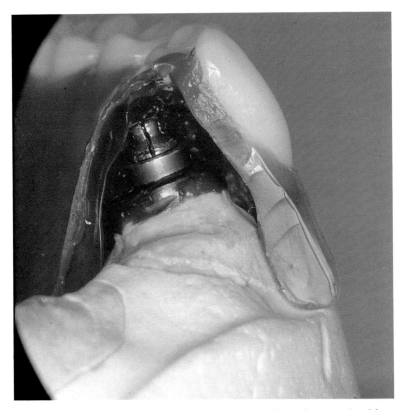

Fig. 5.4 Ceka attachment within a sectioned denture. It can be appreciated how much space is required for this particular attachment and that if it was inclined, either lingually or labially, problems would arise in shaping the lingual contours of the denture base or in aligning the artificial teeth correctly.

The ingenious Kürer Press Stud system occupies rather less vertical space than the Dalla Bona. Secure root anchorage is obtained by a screw-in post which holds the custom-made coping on to the root face (fig. 5.6). This system was the subject of an earlier *British Dental Journal* Teach In and associated booklet.

Where space is particularly restricted, consideration may be given to the resilient Rothermann eccentric attachment (fig. 5.7). The male button-shaped attachment is soldered to the root face coping whilst the clip is embedded in the acrylic resin of the denture. The overall height of this unit is less than 2 mm. Figure 5.8 illustrates the space occupied by this attachment in comparison with the Ceka attachment in figure 5.4.

Fig. 5.5 Diagram of the Dalla Bona 604 attachment.

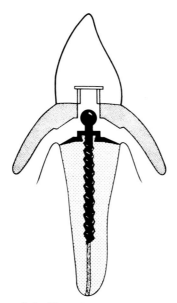

Fig. 5.6 Diagram of the Kürer Press Stud system.

An entirely different approach is adopted in the Zest Anchor system (fig. 5.9), where the nylon male portion is held in the denture and the stainless steel female unit is cemented into a matching cavity cut into the root face by a specially shaped bur. Very little vertical space is occupied by this attachment. The stainless steel insert can be used without a surrounding coping, but great care must be taken to ensure a smooth transition from the edge of the insert to the tooth substance. The

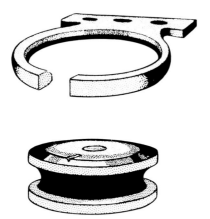

Fig. 5.7 Diagram of the Rothermann eccentric attachment.

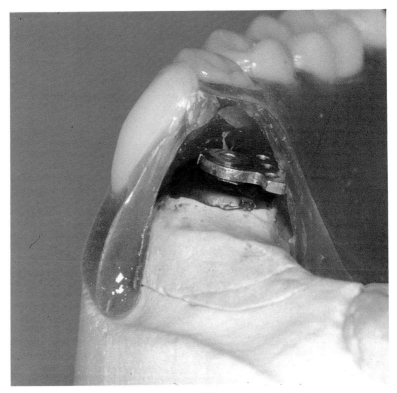

Fig. 5.8 A Rothermann attachment within a sectioned denture.

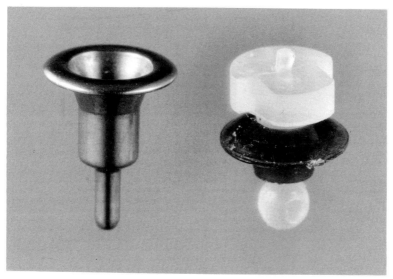

Fig. 5.9 Zest Anchor system.

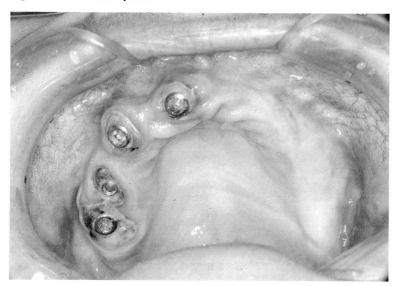

Fig. 5.10 Caries of the root face around Zest Anchor attachments.

surrounding border of unprotected tooth is potentially at risk and root caries is a likely complication unless plaque control is meticulously maintained (fig. 5.10).

The search for positive retention has been tempered by reservations about the complexity and the cost of some of the available precision attachments and several reports have appeared in the literature of simple retentive devices which can be fabricated in the dental laboratory.

Bar attachments

A bar attachment consists of a sleeve, incorporated in the overdenture, which clips over a rigid bar attached to the abutment teeth. A number of the bars are pear-shaped in cross-section thus allowing a degree of rotational movement to occur during function.

The obvious advantage of uniting two abutment teeth via a rigid bar attached to the tooth copings is that the subsequent loading from the denture is shared by the two-tooth unit. It should be recognised, though, that the loading through a bar system is very considerable and therefore the abutment teeth must have particularly sound periodontal attachments.

There are two main disadvantages of the bar attachment: effective plaque control is more difficult to establish and maintain and subsequent maintenance of the overdenture (such as rebasing and replacing the clip units in the denture) can be complicated.

The ideal set of conditions for a bar attachment is one in which the shape of the arch, the positions of the abutment teeth and the shape of the ridge allow the bar to follow the crest of the ridge and ensure sufficient space for the denture, both labially and lingually (fig. 5.11a).

Inevitably, the ideal situation does not always occur and the following problems may become apparent.

1. If the arch shape is pointed in the anterior region or if the abutment teeth are positioned well behind the labial portion of the ridge, a straight bar will pass lingually to the crest of the ridge and cause the overdenture to have excessive lingual bulk (fig. 5.11b). Although it is possible to produce an angled bar, with the clip attachment restricted to the anterior portion, the horizontal forces exerted on the abutment teeth are likely to be excessive (fig. 5.11c).

2. Where there is lack of overall available space in both vertical and antero-posterior directions, the bar may occupy so much of the limited space that it seriously impedes the correct positioning of the anterior teeth.

3. If the roots of the abutment teeth are markedly divergent, it will become impossible to cut post preparations which are of adequate depth and still parallel to each other. This problem can be solved by a particular constructional technique to be described in the next section.

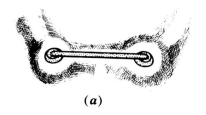

(***a***)

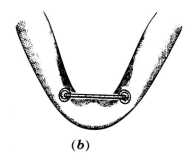

(***b***)

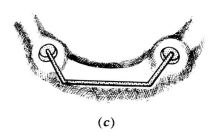

(***c***)

Fig. 5.11 (a) Ideal positioning of abutment teeth and convenient shape of ridge for a bar attachment. (b) Alignment of bar will necessitate excessive lingual bulk of overdenture. (c) An angled bar which is likely to transmit excessive forces to the abutment teeth.

Examples of bar attachment systems

A bar may either be straight or be designed to follow the shape of the ridge. The Dolder bar (fig. 5.12) is a well known and well tried example of straight bar to which the overdenture is attached by a single sleeve. However, it occupies a substantial amount of space.

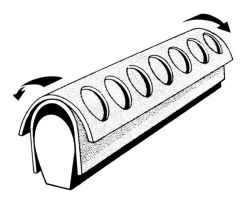

Fig. 5.12 Diagram of Dolder bar and sleeve with perforations for attachment to acrylic denture base.

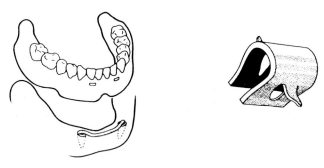

Fig. 5.13 Diagram of Ackermann bar system.

Where an angled bar is used, bent either to follow the shape of the arch in the horizontal plane or the configuration of the crest of the ridge in the vertical plane, attachment of the denture may be obtained by two or more short sleeves. The Ackermann bar is an example of this type of attachment (fig. 5.13).

The basic concept of the Kürer system has been expanded to deal with the problem created by divergent roots. The copings and bar are constructed as one unit and are finally attached to the abutment teeth by coping retaining screws which produce the root anchorage. As these screws are inserted independently of each other, parallelism is not required (fig. 5.14).

Magnets

In recent years, increased interest has been shown in the dental application of rare-earth magnets, particularly the cobalt-samarium magnets.

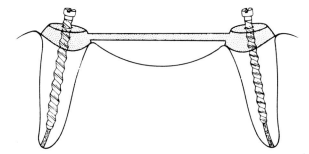

Fig. 5.14 Diagram of a bar attachment uniting abutments with divergent roots.

When compared with other types they have the following advantages: they possess a higher magnetic field strength; they possess a higher intrinsic coercivity (resistance to demagnetisation); their construction allows a close magnetic field to be produced. Magnets do not 'grip' the abutment teeth so strongly. Experiments on photoelastic models, where magnets were attached to osseointegrated implants, showed that there was equitable distribution of functional stresses and that undue loading of the abutment teeth was prevented. Magnets can thus be used in circumstances where the periodontal attachment of the abutment teeth has been reduced. Another point that should be made in this respect is that magnetic retention is self-limiting; once the displacing force exceeds the force of attraction, contact with the abutment tooth is lost and there is minimal possibility of damaging forces being transmitted to the supporting tissues.

One design of magnet which has been used is a reversed split-pole magnet whose ends are protected by a stainless steel shim (fig. 5.15). This magnet, 3 mm high and approximately 5.5 mm long, is contained in the denture and is attached to a thin keeper held in the abutment tooth. Retentive forces within the range 155–980 g have been reported. Some concern was originally expressed that the magnetic units might be prone to corrosion in the oral fluids but recent design changes have overcome the problem. Another type of magnet is illustrated in figure 5.16. This is the Dyna magnet system, in which the abutment teeth are provided with copings in a palladium-cobalt alloy which is receptive to magnetic forces. The small encapsulated cobalt-samarium magnets (2–5 mm high and 4 mm in diameter) are housed within the denture base.

In summary, magnets are simple to use, economical in cost, reduce the force transmitted to the abutment teeth and require little maintenance.

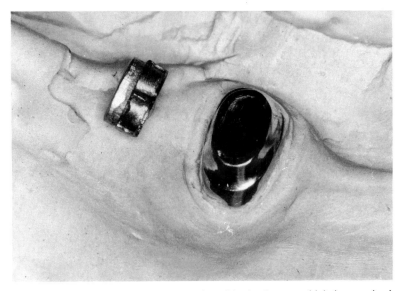

Fig. 5.15 A cobalt-samarium magnet alongside the keeper which is contained within a gold coping on the abutment tooth.

Impression procedures for the overdentures

Once the abutment teeth have been prepared, with or without the provision of copings, impression techniques can follow along conventional lines. The final impression should be taken in a carefully border-moulded special tray so as to ensure that the eventual denture base is stabilised and supported to the optimum degree.

In the absence of marked undercuts a zinc oxide-eugenol impression paste may be used. However, it is occasionally necessary to use an elastic impression material to record an undercut ridge surrounding the abutment teeth. In this case a silicone elastomer is probably the most appropriate material. Needless to say, a spaced tray should be used for alginate impression material.

The problems of recording an impression of tissues of varying compressibility should not be overlooked. These problems are highlighted in cases where the denture-bearing mucosa is particularly compressible. The compressibility of this tissue is, of course, greater than that of the abutment teeth. If $\overline{3|3}$ have been retained, the overdenture is in reality not dissimilar to a bilateral free-end saddle denture (fig. 5.17). If the impression technique results in minimal compression of the denture-bearing mucosa, the denture will tend to pivot about the abutment teeth during occlusal loading. As a result, the teeth will be

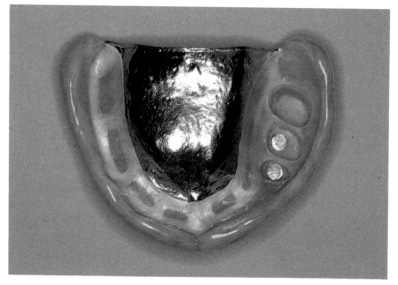

Fig. 5.16 Upper denture with Dyna magnets in |56 region.

called upon to accept additional stresses and the denture-bearing tissues distal to the canines will be loaded unevenly. If, on the other hand, a technique is chosen whereby an impression is obtained of the mucosa in a compressed state, the resulting denture is more likely to load both the mucosa and abutment teeth more evenly during normal function. The objective of the following method is to obtain some compression of the mucosa but minimal displacement of the tooth surface and vulnerable gingival margins.

A close-fitting acrylic special tray is constructed on the study cast and windows are cut in the tray over the abutment teeth and gingival margins (fig. 5.18). The pillars on the tray are conveniently placed for

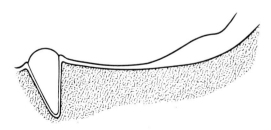

Fig. 5.17 Diagrammatic representation of the lower ridge extending from the canine root abutment to the retromolar pad.

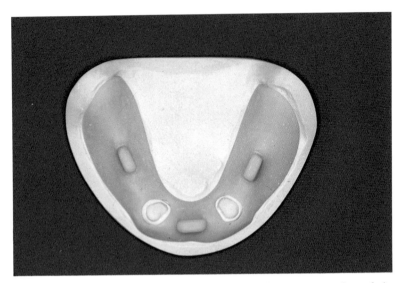

Fig. 5.18 A lower special tray designed to eliminate compression of the abutment teeth and gingival margins.

application of pressure during the impression-taking procedure. After the extension of the borders of the tray has been corrected an impression can be taken with zinc oxide-eugenol impression paste. The viscosity of the impression material ensures compression of the mucosa whilst the excess material makes an easy exit through the windows and therefore does not distort the gingival margins. (If there are marked undercut areas around the ridge a medium-viscosity silicone impression material may be used.) The completed impression is removed from the mouth and the material cleared from around the windows. The tray is then reseated on the mucosa and an impression taken of the exposed abutment teeth and gingival margins with impression plaster.

It is not the purpose of this book to consider details of impression techniques if attachments are used, as these techniques are well covered in specialised texts. It is, though, appropriate to make the point that the principles discussed earlier in this section should be applied equally to overdentures retained by attachments.

If stud attachments have been prepared for the abutment teeth, the objective of the final impression technique is to ensure that the attachment, together with the die on which it was constructed, can be located accurately on the master cast. The usual technique is to employ a transfer dummy which is seated on the abutment tooth before the final impression is recorded. The dummy projects through the windows of the special tray and, after the overall impression has been taken, it is firmly fixed to the tray with either cold-curing acrylic resin or impression plaster.

A similar approach is adopted for the bar attachment to ensure that the dies of the abutment teeth are correctly related to the remainder of the denture-bearing tissues. Once the cast has been poured, with dies in position, the attachments are constructed.

Discussion

In concluding this chapter it is worthwhile to consider some of the evidence on the effect of copings and attachments on the health of the abutment teeth. With respect to stress distribution, current knowledge is the result of studies on photoelastic models.

The ideal arrangement of stress distribution is that the occlusal forces from the overdenture are transmitted through the long axis of the abutment teeth with minimal horizontal torque. A dome-shaped coping appears to transmit the least amount of stress down the long axis of the abutment tooth with rather more of the stress being transmitted widely over the edentulous ridge. The more positive the attachment, the higher is the stress to the abutment teeth. The more rigid designs of bar connector transmit higher stress to the abutment teeth and cause more torquing forces.

More details of the long-term clinical evaluation of attachments will be presented in the next chapter. The point should perhaps be made at this stage that the health of abutment teeth is more likely to be prejudiced by the presence of attachments.

In view of the cost of attachments and the complications that may arise during their fitting and subsequent maintenance, the authors believe that a simple dome-shaped abutment (with or without coping) should initially be employed in the majority of cases. Options still remain open and it is possible to progress to attachments if necessary.

6

Clinical Evaluation and Maintenance

Clinical evaluation of overdentures

This chapter considers the likely post-insertion tissue changes associated with overdentures and the measures that can be taken to minimise complications.

There have been a number of follow-up studies in relation to partial dentures and it has been clearly and convincingly demonstrated that good oral hygiene is the single most important factor which influences the continuing health of the supporting tissues. The chance of damage to the periodontal tissues is increased with inadequate oral hygiene and poor prosthetic design. The evidence to date suggests that these conclusions are equally applicable to overdentures (fig. 6.1) although there is very little documented evidence on long-term problems associated with overdentures. Only a few studies are available on the clinical evaluation of overdentures and most have concentrated on periods up to 5 years post-insertion.

One study was concerned with the immediate post-insertion period but involved only a small number of patients; no significant differences were found in the pocket depths or mobility of the abutment teeth up to 8 months after placement of the overdentures. In a further study of 50 patients, carried out between 6 and 47 months after provision of overdentures, eight of the patients exhibited caries of the abutment teeth. The periodontal health of 30% of the patients was rated as 'failing' and of a further 28% as 'fair'. These observations led to the comment that an exceedingly high incidence of periodontal pathology is the major problem in the long-term use of overdentures. In contrast, another study reported little change in gingival health over a 2-year period although the initial condition of some of the teeth had been less than optimal.

Another review of 17 patients who had worn overdentures for between 4 and 25 months, demonstrated that when fluoride was applied

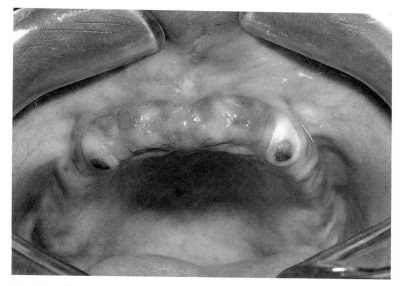

Fig. 6.1 3|3 which have served as root abutments under a complete upper overdenture for 5 years with no deterioration in the condition of tooth structure or gingival health.

daily by placing a drop of gel in the root face depressions of the overdentures, only 5% of the teeth developed caries. Over 20% of the non-fluoridated abutments suffered carious attack.

A much larger study examined the caries prevalence of 210 abutment teeth after one year; 35% of these teeth had become carious. The caries often developed within a short period of time after insertion of the overdenture — as little as 2 months in some cases. After a further year, a significant decrease in caries from 35% to 19% was noted when patients were placed on a regime of 1% neutral pH sodium fluoride solution, used daily. There was also a comment on the effectiveness of silver nitrate, although many patients objected to the discoloration of the teeth and dentures.

Two studies, of 5 and 4 years' duration respectively, have reported a caries prevalence of 13.6% and 22%. A level of 35.7% was recorded in a third trial, of 3 years' duration, consisting of only seven patients. A recent 10-year follow-up study found that of the 77 overdenture abutments that were present when the overdentures were originally placed, 66 were still present. Seven of the abutments were lost because of periodontal disease. Caries were detected on eight of the 66 teeth still present. None of the patients in whom caries were detected was using fluoride on a regular basis. The study therefore reported a caries prevalence of 19%.

An American study of 254 patients treated over a 12-year period identified the rate of tooth loss of overdenture abutments as only 4.2%, with caries and periodontal disease being the main causes of failure. Although 667 abutments were assessed, since the treatment had been provided over an extended period, some of the cases could not have been followed up for more than 2 or 3 years. The comment was made that most of the failures could have been prevented if there had been improved communication between dentist and patient in order to stress the importance of effective plaque control and the need for regular recall.

It is also important to be on the look-out for changes in social circumstances which may interfere with previously effective home care. Prolonged illness, drug therapy reducing salivary flow, hospitalisation or death of a spouse are examples which are, of course, likely to occur more frequently amongst older people.

There have also been studies of tissue health under complete overdentures retained by attachments. In a group of 31 patients examined between 6 months and 4 years after fitting overdentures, only 10% of the abutment teeth were surrounded by healthy gingivae.

There was no difference in the prevalence of gingivitis between the wearers of the newest and oldest dentures. There was an associated deepening of the gingival pockets around 35% of these abutments. Retraction of the gingival margins was evident in all cases where the dentures were older than 3 years, but in only 28% where the dentures were less than 1 year old. Thirty-nine per cent of all abutment teeth were carious, but since the root faces were covered by copings and attachments the attack was only evident in those teeth with retraction of the gingival margins.

In the majority of cases the dentures were worn at night, although all the patients had been instructed otherwise. Only 50% of the dentures were plaque-free, the areas in the vicinity of the attachments being most susceptible to plaque formation.

A review of 12 patients with bar attachments concluded that there was an increased prevalence of gingivitis and deepened pockets around the abutment teeth and that these findings could be attributed to inadequate oral hygiene. Unsatisfactory oral hygiene was found in 9 of the 12 patients but over the 3-year observation period there was no marked loss of periodontal support. However, the likelihood of the poor gingival condition leading to deterioration of the periodontal

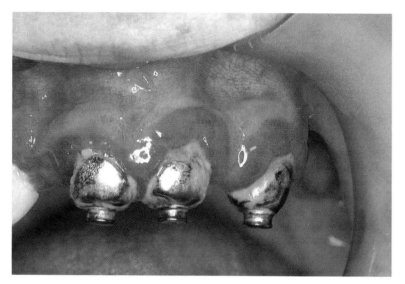

Fig. 6.2 Gross plaque accumulation leading to inflammation and gingival hyperplasia around three maxillary anterior abutments which incorporate Rothermann eccentric attachments.

supporting tissues over a longer time interval could not be discounted. In a larger review of 110 bar attachment dentures it was concluded that the degree of gingival inflammation of the supporting tissues was directly related to poor hygiene procedures.

A number of studies have also produced evidence on the prevalence of fracture of overdentures. It appears that overdentures retained by attachments are more liable to fracture. It is also apparent that because of resorption of bone in areas distant from the abutment teeth there is loss of denture stability with the risk of uneven loading of the abutment teeth and edentulous ridges.

In summarising the available information on tissue reaction it is apparent that there are two major problems associated with wearing overdentures: first, periodontal damage, which may take the form of gingival hyperplasia or retraction of the gingival tissues together with a loss of bony support (figs 6.2 and 6.3); second, caries of the root face or, if retraction of the gingival margin has occurred, of the root itself (fig. 6.4).

A definite relationship has been established between the level of plaque control and periodontal damage and it has been shown that there is a special risk of periodontal deterioration if attachments are used. The level of risk can be substantially reduced by emphasising the importance

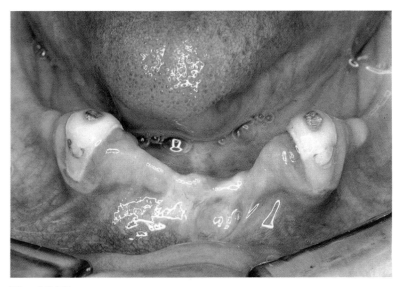

Fig. 6.3 $\overline{3|3}$ abutments showing retraction of the gingival tissues together with some inflammatory change affecting the marginal gingivae.

of preventive measures, by instruction in oral hygiene procedures, by regular inspection and maintenance of the dentures, and by the application of topical fluoride.

Preventive measures
Plaque control of the abutment teeth
Maintenance of periodontal health is one of the key factors in the long-term success of overdentures and it will depend upon effective plaque control by the patient.

The plaque can more easily and effectively be removed with the toothbrush if the abutment teeth are separated. Since many patients might not consider brushing root surfaces to be part of the normal oral hygiene programme, it needs to be demonstrated and continually reinforced at follow-up visits. A disclosing solution should be used to demonstrate the presence of plaque on the abutments. There is evidence that chlorhexidine can be helpful in the control of plaque and that it acts as an antimicrobial agent. It can be delivered to the abutment teeth either as a mouthrinse or in gel form. In the case of overdenture abutments it is likely that it will be most effective if used in small quantities in gel form when placed in the depressions of the impression surface of the denture

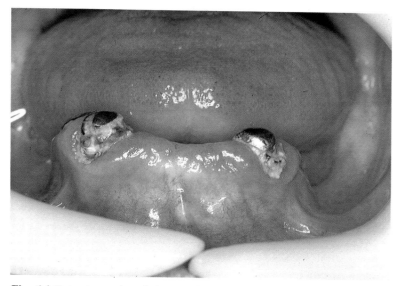

Fig. 6.4 Extensive caries of the root face on two lower canine abutments.

which relate to the underlying abutments. It can be used this way once or twice daily on a regular basis although some staining will occur on the abutments. Some might feel that its use should be reserved for 'at risk' individuals. A fluoride toothpaste should be recommended to enhance caries inhibition. Early scientific evidence suggests that fluoride and chlorhexidine could be combined for optimum benefit and effect.

In an attempt to improve oral hygiene standards a toothbrush specially designed and shaped for overdenture abutments might be considered. One recently described has outer rows of bristles which are three times as long as the inner rows. It is claimed that this allows a simple horizontal scrubbing movement of the handle and that it is easier to use and more efficient.

The findings of a recent 3-year follow-up of 35 older patients show that initial non-surgical periodontal treatment, together with good advice and motivation, can improve the periodontal condition and allow that improvement to be maintained as long as regular recall visits are kept up. An additional point is made that initial oral hygiene status is a poor predictor of the patient's eventual co-operation and of long-term prognosis. In other words the individual patient must be given an opportunity to respond to professional advice.

Plaque control of the denture

As denture hygiene is often neglected by the patient, a regular and efficient denture cleaning programme is essential to maintain the health of the supporting tissues. Even when an effort is made by the patient, the result is often poor due to either incorrect brushing technique or to the shape of the impression surface of the overdenture where the presence of undercuts makes it extremely difficult to use a brush efficiently. Furthermore, not all denture-cleaning solutions are effective substitutes.

The overdenture should be brushed after meals with a soft toothbrush, using soap and water to remove loose deposits. Particular attention should be paid to the impression surface in the region of the root face depressions. As with the abutment teeth the effectiveness of the cleaning procedure can be monitored by using a disclosing solution on the denture surface. Periodontal health is more likely to be maintained if the patient can be persuaded to leave the overdenture out whilst sleeping.

Clinical trials of the effectiveness of denture cleansers in removing plaque show hypochlorite solutions to be particularly efficient. The long immersion period recommended by the manufacturers has the added advantage of ensuring the denture is not worn overnight. It has been reported that the hypochlorites are harmful to metal components of dentures, but no ill-effects should be encountered provided a buffered hypochlorite is used and the recommended short immersion period adopted.

Particular care should be taken in cleaning precision attachments housed in the denture base, as plaque is more likely to collect in the relatively inaccessible areas.

Care of the root face

Compared to enamel, dentine is much more susceptible to carious attack. Rigorous measures are required to prevent the eventual decay of the abutments, particularly if the root face is not protected with a gold coping. The use of fluoride, in various forms, has been shown to be the most efficient of the inhibitory agents in the prevention of dentine caries.

Topical fluoride applications can be carried out using acidulated phosphate fluoride gel immediately after the root face preparations are completed and at subsequent follow-up appointments. The root faces should be polished and inspected before fluoride application. A programme of self-application of fluoride by the patient is also recommended for maximum benefit. However, it should be appreciated that patient compliance with self-medication is often suspect and lapses will occur unless there is thorough understanding of its importance and consequences. A regime that can be recommended is the daily application of acidulated phosphate fluoride gel drops applied in those depressions in the impression surface of the denture which overly the abutments. Fluoride gels are available commercially which stay within the safe recommended dosage when applied in this way.

An alternative is the daily use of a suitably diluted sodium fluoride mouthwash. The abutments are rinsed thoroughly and the solution spat out. This avoids possible ingestion of excessive amounts of fluoride which could occur with the over-enthusiastic home application of fluoride gel.

Maintenance
The abutment teeth and periodontal tissues
Maintenance of gingival health has already been stressed since it is one of the key features in the long-term success of overdentures. Thus, at each recall particular care should be taken to prevent the onset of periodontal disease. A check should be made on plaque accumulation (fig. 6.5), both on the tissues and on the denture, and oral hygiene instruction should be reinforced. It should be remembered that, in the long term, more active periodontal treatment may be required to re-establish tissue health.

If there is gingival recession, it may be necessary to reshape the dome preparation, in which case the impression surface of the denture must be modified to maintain a close fit.

If the root face has become carious, it will need to be modified and possibly a gold coping provided (fig. 6.6). However, it is unwise to fit gold copings much earlier than 6 months after inserting the denture in case further gingival recession occurs (figs 6.7 and 6.8).

The denture
Immediate overdentures
If an immediate overdenture is fitted, early review visits will be required as with any immediate denture; that is, the following day, at 1 week, at 1 month and at 3 months after extraction. At these visits the progress of healing is followed and the cause of any mucosal irritation is diagnosed and treated.

The occlusion should always be assessed and corrected where necessary, while any deterioration in fit is treated either by using a

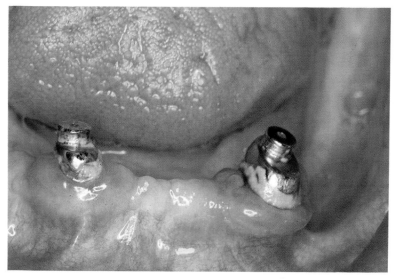

Fig. 6.5 Plaque accumulation and caries affecting two mandibular abutments which incorporate Rothermann eccentric attachments.

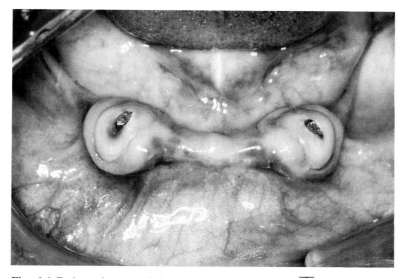

Fig. 6.6 Early caries around the amalgam restorations in $\overline{3|3}$ abutments. The restorations require to be enlarged and replaced. Gold copings are not necessary at this stage.

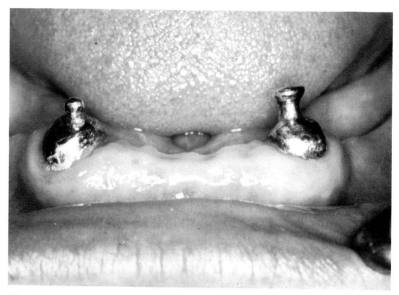

Fig. 6.7 Gold copings with simple attachments on $\overline{3|3}$ abutments shortly after cementation.

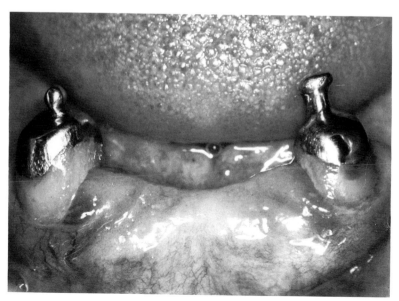

Fig. 6.8 Considerable gingival recession has taken place. The preparation is now more susceptible to caries and periodontal disease and to unfavourable loading from the denture.

conventional relining procedure or by using a temporary soft lining material. If a temporary soft lining material is used, particular care should be taken to maintain a plaque-free surface.

Some combinations of denture cleansers and temporary soft lining materials are incompatible; laboratory studies have shown that one of the soft materials (ViscoGel: De Trey Division, Dentsply Ltd, Weybridge, Surrey) is compatible with a commercially available hypochlorite denture cleanser (Dentural: Farillon Ltd, Romford, Essex). In any case, the temporary material should be replaced by heat-curing acrylic resin as soon as possible to avoid potential irritation of the marginal gingivae.

Review appointments at 6-monthly intervals are usually sufficient when the initial maintenance phase is completed.

Impression surface
Bone around the abutment teeth (or implants) is likely to be retained, while resorption is more likely to occur at some distance from the abutments. As a consequence, an antero-posterior rock of the denture is likely to develop if the abutments are located only in the anterior region. This effect is more likely to occur in the lower jaw. Stability of the denture is re-established by relining, using either a conventional technique or one of the materials which have recently become available for chairside use.

Conventional technique. As with all relining techniques it is necessary first to check the occlusion, to correct any border extension deficiency and to remove undercuts from the impression surface of the denture before proceeding with the impression. In the absence of marked undercuts around the abutment teeth, zinc oxide-eugenol impression paste can be used. If undercuts exist, then a silicone elastomer is the impression material of choice.

If it is considered that an impression of the edentulous areas should be recorded under some degree of load (with, at the same time, minimal compression over the abutment teeth), vent holes can be cut in the denture over the abutment teeth to allow excess impression material to escape.

The relining of dentures involving bar or stud attachments can be a difficult and involved procedure and errors in the localisation of the attachments may occur. This may result in a denture that will reseat without the attachments engaging correctly. The reader is well advised to be aware of the problem and to consult the manufacturer's instruction manual before undertaking such a procedure.

Chairside technique. Provided that no substantial undercuts exist in the denture-bearing areas or around the abutment teeth, an autopolymerising poly (butyl methacrylate) acrylic resin (Total: Staident International, Staines, Middlesex), suitable for intra-oral use, can be used to perform a chairside reline. The impression surface must be thoroughly cleaned and dried and the recommended bonding agent applied. The resin is placed in the denture before proceeding with the impression. After approximately 3 minutes the denture should be removed and the impression surface checked. If satisfactory it may be returned to the mouth to allow complete curing of the material. A potential disadvantage of this technique is the resultant increase in thickness of the denture base. While this bulk is of little consequence in the lower denture, provided the dimension of the freeway space remains within acceptable limits, increased thickness of the palate is normally unacceptable.

Occlusal surface
It is essential to maintain a correctly balanced occlusion to ensure stable dentures and an even distribution of occlusal load so that denture comfort is enhanced and the health of the underlying tissue sustained.

Occlusal wear of acrylic resin teeth usually occurs gradually. However, the rate of wear is increased when an overdenture occludes with natural teeth in the opposing jaw. The increased stability and support of overdentures also allow greater loads to be applied during chewing. A recent 10-year report of a longitudinal recall of 28 overdenture patients demonstrated that although the occlusion of 22 patients was still rated as good the majority of the 23 patients with acrylic resin teeth showed moderate to severe wear. As you would expect, those with porcelain teeth had the least amount of wear. If the dentures are generally satisfactory, except for an occlusal imbalance caused by wear of the posterior teeth, it is possible to replace the teeth without remaking the dentures. An occlusal registration is obtained with the mandible in the retruded position and with the teeth slightly apart to ensure that deflecting occlusal contacts are not made. The dentures are then mounted on an articulator so that the old teeth can be replaced. Although this procedure is simple, it does have the disadvantage that the patient is without the dentures while the laboratory work is being carried out.

Postscript
This chapter has highlighted the various forms of damage that can occur under overdentures. It should be remembered, however, that with a

well-motivated patient and with careful maintenance of the prosthesis, the risks can be minimised. Bearing in mind the overall advantages of overdenture treatment and the benefits it can give to certain groups of patients, the recent interest in the technique is fully justified and is likely to develop further.

7

An Alternative Approach: Implant Retained Overdentures

The previous chapters described the clinical indications and uses of overdentures which utilise natural teeth as abutments. The next three chapters will discuss the use of dental implants as an alternative to natural abutments. This chapter describes the difficulties faced by the edentulous patient in adapting to dentures. It reviews the criteria for patient selection and the indications for using implants to support overdentures.

Introduction

For some patients the provision of overdentures is the first step on the way to the eventual wearing of complete dentures. One of the aims of overdentures is to aid in the adaptation to complete dentures. Society is however changing its views on complete dentures and it is becoming apparent that many patients find the thought of wearing complete dentures daunting or even unacceptable. There is now an alternative approach that may prove an effective treatment of the edentulous condition. This has arisen following the development of artificial root forms which can be successfully implanted into the jaw. These implants are machined to accept attachments used to secure a denture. If a sudden or even expected failure of a natural root abutment occurs, it should be possible to consider the replacement of this tooth with an implant after an appropriate healing period, thus extending the life of the overdenture. Dental implants can also recreate the overdenture situation in an edentulous patient. The added advantage here is conservation of the remaining alveolar bone brought about as a direct result of implant placement.

Problems associated with complete dentures

The complete loss of the natural dentition is still a relatively common occurrence in the Western world. In the United Kingdom it has been estimated that 8 million adults are wearing complete dentures. Since it is so common the majority of people accept the condition as a natural process of ageing and it attracts little in the way of sympathy. This situation is due in part to the very successful restoration of the missing dentition by removable dentures.

Despite this the loss of the remaining natural teeth imposes a severe handicap for many people. There is an obvious change to the shape of the lower third of the face with loss of lip support which can markedly age the patient. A major impact of tooth loss is the concurrent loss of the supporting alveolar bone. If this loss is severe it will dramatically affect the ability of individuals to control their dentures and it is well known that masticatory ability and bite force are dramatically reduced after the loss of the natural dentition. Diet may become severely restricted and this may inhibit social activities. The movement of the denture on thin atrophic mucosa can result in ulceration and pain as well as causing speech difficulties.

There are psychological implications associated with tooth loss. These can go back to childhood experiences, for example memories of seeing their parents not wearing dentures, or the loss of teeth due to trauma. Finally peer group and family attitudes to the person wearing dentures can be particularly important. The modern day media tend to highlight the 'perfect body' and removable dentures do not feature in this image. The wearing of dentures is often the subject of comedy and is seen by many to symbolise ageing and a lack of vigour. It is a source of surprise to many clinicians how well the majority of patients adapt to dentures.

A combination of factors therefore results in a small group of patients for whom the wearing of any type of denture is intolerable. The traditional approach to such patients has been to advise the use of specialised prosthetic techniques or to prescribe a preprosthetic surgical procedure.

Special prosthetic techniques for the edentulous patient
The grossly resorbed alveolar ridge has always posed problems in the provision of stable, comfortable and retentive dentures (fig. 7.1). The traditional approach has been to obtain accurate impressions of the maximum denture bearing area while avoiding overextension in areas of active muscle attachment. Efforts have been made to improve muscular control of the denture by recording volumetric impressions of the denture space. For maximum retention it is thought that a denture should fill all the space between the tongue on one side and the cheek

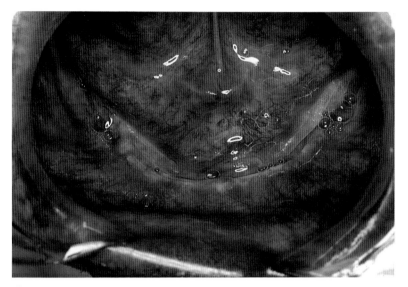

Fig. 7.1 Gross atrophy of the mandibular ridge.

and lips on the other. This zone of minimal muscular conflict is known
as the neutral zone and the method used to record this is the reciprocal
impression technique. This technique is used to shape the polished
surface of the denture and can act as a guide to the positioning of the
teeth. The method of recording neutral zone impressions is very
technique sensitive and depends to a large extent on the material used
for the impression.

Other prosthetic techniques have aimed to reduce the loads a patient
may apply to a resorbed atrophic ridge. This can be achieved by reducing
the number and width of posterior teeth and using a resilient lining on
the impression surface. Success with this approach will depend on a
number of factors including the individual dentist's clinical skills, high
quality technical work, patient cooperation and the health of the
underlying denture bearing tissues.

Preprosthetic surgery as an aid to complete denture wearing
The objectives of preprosthetic surgery have been to enlarge the
available denture bearing area and to reduce the displacing effects of
adjacent musculature. There are many texts describing preprosthetic
surgical techniques and only a very brief review of one or two procedures will be detailed here. The results of major preprosthetic surgery
have often been disappointing and associated with significant compli-

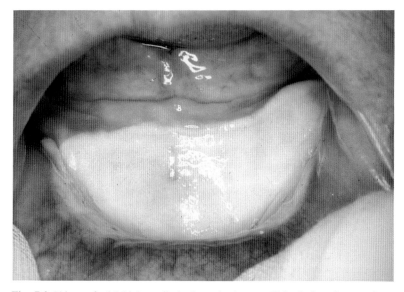

Fig. 7.2 Skin grafted labial vestibuloplasty in the mandible designed to produce a large increase in denture bearing area.

cations. In recent years surgical enlargement of the denture bearing area has been achieved by two procedures; vestibuloplasty and ridge augmentation.

Vestibuloplasty

This procedure is indicated if a patient has enough basal bone remaining (greater than 15 mm total ridge height) but lacks sulcus depth. The surgery relies on creating a larger denture bearing area by detaching the origins of muscles on the labial and or lingual sides of the residual ridge. These muscle attachments would have been responsible for displacing the denture prior to surgery. The wound heals by secondary epithelialisation or with the aid of skin or mucosal grafting (fig. 7.2). The success rate in terms of patients able to wear dentures in comfort has been shown to be about 70-80%.

The procedure is associated with complications including damage to the cutaneous sensation of the lips and chin (30%) and a 25% incidence of adverse changes to the lower third facial profile due to a prolapse of the soft tissues of the chin. Following surgery the patients are able to apply greater pressure to the residual alveolar ridge during mastication and this in turn may result in an enhanced rate of alveolar ridge resorption.

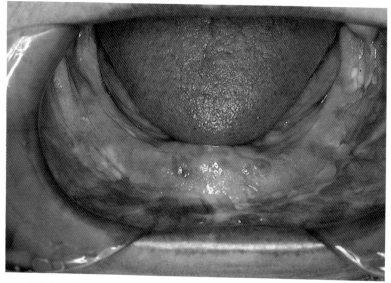

Fig. 7.3 Mandibular ridge following sandwich augmentation and subsequent skin grafting. Denture movement has produced areas of frictional keratosis.

Ridge augmentation

The aim of this surgery is to increase height and width of the residual alveolar ridge using bone grafting procedures. The operative technique is more extensive than vestibuloplasties as it involves soft and hard tissue. Some patients will be unsuitable because of age, health or ability to cope with surgery. Many augmentation procedures recommend that a vestibuloplasty be undertaken 6 months later to take full advantage of the surgically created ridge (fig. 7.3). It has been clearly demonstrated that interpositional bone grafts and pedicled bone grafts are superior to onlay techniques. Even so the prosthetist can still be faced with a loss of 50% of the surgically created ridge over a 3-year period. This, combined with a poor quality denture bearing tissue stretched over the bone graft, can make it very difficult to provide satisfactory dentures. With more major surgery there is an increased risk of morbidity and there have been reports of an 80% incidence of mental nerve paraesthesia.

Recently there has been a great deal of interest in the polycrystalline ceramic, hydroxylapatite. Its physical and chemical characteristics are similar to cortical bone. It can be formed into various ridge shapes and has been positioned between bone and mucoperiosteum. It has been used for filling extraction sockets or onlay augmentations of atrophic ridges using particulate or porous block forms. It has also been used with

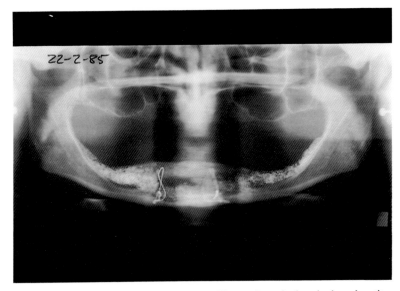

Fig. 7.4 Ridge augmentation using bone chips and particulate hydroxylapatite.

bone chips in pedicled bone graft osteotomies in an attempt to reduce post-surgery ridge resorption (fig. 7.4). Used in this way there is still some doubt about the ability of bone to bond to hydroxylapatite. Failure to integrate could result in migration of the hydroxylapatite particles to form secondary ridges which may cause a problem for final denture construction.

These surgical solutions to difficult prosthetic problems still require the patient to wear a removable denture. Success rates of between 70 and 80% in terms of patient satisfaction have been claimed in the short term. There is still doubt about the long term success and durability of the surgical result. The final ridge shape is not predictable and in some patients it could lead to a worsening of their situation. The last decade has seen a dramatic reduction in the use of major preprosthetic surgery except as an aid to the placement of dental implants.

The use of implants in the edentulous patient

In marked contrast to preprosthetic surgery, dental implants have developed rapidly in the last 20 years from a form of treatment that was used as a last resort to one that is recommended routinely for many patients. This has been based on rapidly expanding technology which has improved implant design. Concurrent with this has been an improved surgical technique for the placement of implants. Much of the

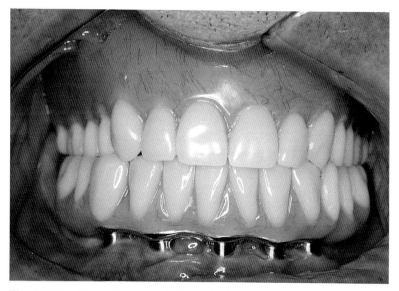

Fig. 7.5 Mandibular fixed prosthesis retained on five implants.

progress in this area has been due to the developmental work of Brånemark and his colleagues. This Swedish group reported on a series of well controlled clinical trials evaluating titanium implants over a 10–15 year period and produced the long awaited evidence that artificial tooth root implants could be placed with a predictable chance of success.

As has already been described earlier in this book, attempts to retain two or more teeth for use as abutments for overdentures have played a major part in prosthetic treatment. Dental implants make excellent substitutes for roots and can easily be used to support and retain overdentures. The type and number of implants used in the edentulous jaw will determine whether a fixed prosthesis (fig. 7.5) or an overdenture (fig. 7.6) is to be used in the final restoration.

The fixed prosthesis in the edentulous patient requires enough bone to support four to six implants. A minimum total ridge height of 15 mm is recommended. This type of restoration is particularly useful in patients who have never accepted the loss of their natural teeth and find dentures intolerable. As the prosthesis is totally supported by the implants there is no need to load atrophic denture bearing mucosa. Patients have reported a dramatic improvement in their masticatory ability following this type of restoration. A fixed prosthesis is valuable where there are marked anatomical problems present which have interfered with the fit of dentures, such as uneven 'knife like' ridges.

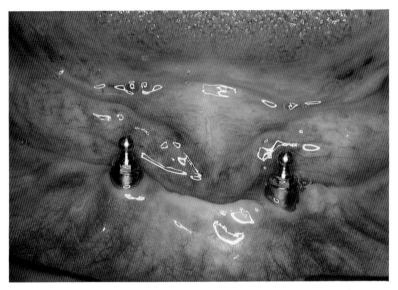

Fig. 7.6 Two O-Ring abutments which support a mandibular overdenture.

The actual number and size of implants used will be determined by the bone quality and quantity. Although the minimum number of implants required for a fixed prosthesis is four, normally five or six are placed to allow for possible implant loss. The shape of the arch is important in the prosthesis design. The implants should be placed in an arc so that there is at least 10 mm distance between the most anterior and the distal implant (fig. 7.7). The maximum length cantilever designed for a curved arrangement of the implants would be twice this anterior/posterior measurement in the mandible. The extent of the cantilever is somewhat reduced in the maxilla due to the poorer bone quality. The length of the cantilever should also be reduced if short implants have been used. As well as requiring more bone for implant placement this type of restoration does not always give adequate soft tissue support. The gap left between the prosthesis and the underlying mucosa can allow the passage of fluid or even small particles of food to escape from the mouth. Speech may also be affected particularly in maxillary cases. The surgery is more prolonged and the prosthetic and technical work more demanding, which all leads to increased costs. Finally the appropriate level of maintenance required from both patient and dentist is quite time consuming.

The implant retained overdenture usually requires only two implants although often three or more may be placed to give extra retention and a fall back position should an implant fail. This approach is

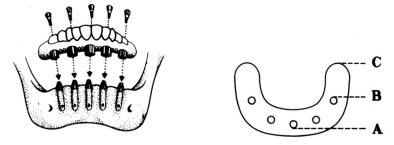

Fig. 7.7 The length of the cantilever B–C in a fixed prosthesis should not be more than twice the distance between the most anterior and posterior implant A–B.

commonly used in the very atrophic ridge where there is less than 15 mm of total ridge height left. The surgery, prosthetic and technical work is much simpler and more economical than for the fixed prosthesis. An overdenture can give good support to the surrounding soft tissues and should not result in any speech or saliva problems. Some patients find the thought of cleaning multiple implants supporting a fixed prosthesis daunting. For these a removable denture is much easier to cope with. It is occasionally necessary to consider converting a fixed prosthesis to an overdenture. This can occur if an implant should fail or if the patient through illness or for some other reason becomes unable to maintain a fixed prosthesis. An overdenture can also be used as an interim prosthesis should a fixed prosthesis require removal for repair at any stage.

The only disadvantages of overdentures relate to those patients who are completely opposed to wearing removable dentures. Some overdentures may be bulky, especially over the attachments. This difficulty of accommodating the attachments may lead to speech problems or to an increased risk of denture fracture because of reduced denture base material bulk. The attachments and denture will require regular maintenance. Implant placement errors or positional problems due to the shape of the residual alveolar bone may result in divergent implants which may lead to aesthetic, speech and retention problems. Finally the functional improvement in terms of bite force and masticatory ability may be less with an overdenture than a fixed prosthesis. Clearly, however, the improvement provided by an overdenture over the edentulous state is still very considerable.

Implant types

Dental implants can be classified by the materials used in their construction and by design.

Materials

There are many materials used in the manufacture of dental implants today. The majority are made from metals such as 314 stainless steel, titanium, titanium alloy and cobalt chromium alloys to name just a few. There are also a growing number of non-metals, for example porcelain, high density aluminium oxide, ceramic, sapphire and carbon. Many different shapes, contours and surface textures (smooth or roughened) are used with porous and non-porous properties. Non-metal coatings, for example tricalcium phosphate or hydroxylapatite, have been applied to different implant designs. The chemical and physical characteristics of an implant surface may be crucial to the adhesion and migration of cells and may influence the adsorption of various serum proteins.

These materials can be further subdivided into two groups according to the reaction they provoke in the surrounding bony tissue. In one group the implant does not influence the adjacent tissues and is termed 'bioinert'. In the second group the implant surface appears to bond to the living bone and this has been described as a 'bioactive' reaction. Hydroxylapatite and tricalcium phosphate are the main representatives of the bioactive implants.

Design

The simplest and best known classification of dental implants relates to an implant's design and position in the jaw.

Subperiosteal implants

The subperiosteal implant was developed in the 1940s and was designed for the atrophic edentulous ridge. It is more commonly used in the mandible and consists of a framework resting on the surface of the alveolar ridge (fig. 7.8). In general it is not claimed that this implant integrates with the underlying bone. The early designs were quite crude with various materials being used in their construction (cobalt chromium, steel and titanium alloy). The implant usually supported an overdenture-type of appliance.

The construction of the framework has been aided in recent years by the use of computer generated models of the jaw. This has made it possible to eliminate the surgical stage of impression taking. The framework has also been coated with hydroxylapatite in an effort to improve its success rate. Today its main recommendation for use would be in those situations where not enough alveolar bone remained to accommodate an endosseous implant (residual bone height of less than 8 mm). Some workers have reported a success rate of 54% over a 14-year

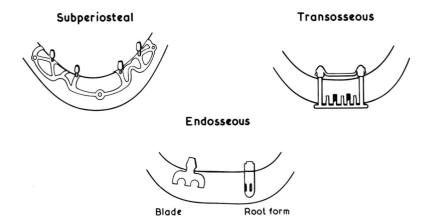

Subperiosteal   **Transosseous**

Endosseous

Blade   Root form

Fig. 7.8 The main types of dental implants in common use today.

period. Many other studies have shown a less promising outcome. It is now felt that although the initial success rate is high, this type of implant will fail eventually. Failure when it occurs results in marked damage to the residual alveolar ridge and denture bearing tissues.

Transosseous implants
These implants are used exclusively in the mandible and there are two main types namely the mandibular staple bone plate and the Bosker transmandibular implant (fig. 7.8). Both types are inserted through the mandible from its lower border and exit into the mouth. To place them usually requires an external incision in the submental region and a high degree of surgical skill. Five to seven parallel holes are prepared in the base of the mandible, two of which penetrate the mandible and exit into the mouth. These two transosteal pins support an overdenture. The mandible must have at least 10 mm of bone height remaining. The main drawback of this implant is that it is restricted to the mandible, needs extensive surgery when placed and is complicated to repair if the transosseous element should fracture. A recent publication of a 16-year evaluation of the mandibular staple bone plate claimed a 90% success rate for periods in excess of 8 years.

Endosseous implants
This group represents one of the most common forms of implants in use today. Typically they are designed as a screw, pin or blade and have been made from materials such as gold, cobalt chromium, carbon, ceramic, stainless steel and titanium.

In the late 1950s blade vent implants were introduced. As their name suggests these are flat but various designs and materials have been used. The Linkow blade seems to have been one of the most popular (fig. 7.8). A one-step procedure was used for the placement of this implant through the mucosa into bone. The early implants were manufactured from chromium and nickel. Currently titanium alloys, sometimes coated with hydroxylapatite, are used. They are most often used for the partially edentulous patient usually replacing missing posterior teeth. Although these implants were quite common their use has been criticised by various workers. Complications ranged from various soft tissue problems and bone loss around the implant to a high failure rate of around 40-50% over 10 years.

Root form implants
There is a multitude of this type of implant design (fig. 7.9). Some are conical, simulating a root in shape; others are cylindrical with or without screw threads or perforations. The majority are made of metal including commercially pure titanium, titanium alloy, chrome nickel vanadium or steel. They also may be coated with other materials such as plasma sprayed titanium, tricalcium phosphate or hydroxylapatite. The effect of all these various coatings on the long term survival of the implant has not yet been fully evaluated.

The techniques used to place these implants vary considerably. Some are installed in a single stage procedure and are immediately exposed into the mouth. Others are placed in a two stage procedure in which case at installation the implant is buried in the bone and the mucosa closed, isolating the implant from the oral environment for an initial healing period of 3 months in the mandible and 6 months in the maxilla. The implant is then exposed at a second procedure and a transmucosal abutment is fitted. Implants are usually inserted into intact bone. However some types can be placed in fresh extraction sites providing there is no local pathology. The Tubingen implant is an example of one suitable for fresh extraction sites. The implant is shaped like an irregular conical cylinder with surface lacunae that are claimed to allow for osteocytic ingrowth. This implant is made from aluminium oxide and is therefore brittle and may not be suitable where high lateral loading forces are expected.

Of the two stage metal screw implants, the Brånemark design has the most documented research available. Detailed laboratory investigations as well as longitudinal clinical studies which have shown the system to have a high success rate with little associated morbidity. A success rate

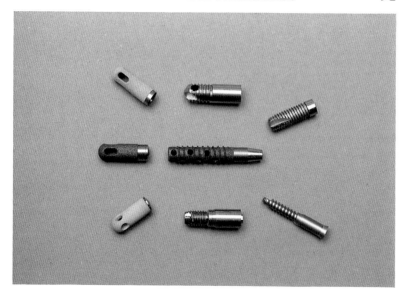

Fig. 7.9 A selection of root form dental implants with various design features.

of 81% in the maxilla and 91% in the mandible in terms of functioning implants over 15 years has been reported. This high success rate is thought to be due to the use of commercially pure titanium which develops an oxide layer on exposure to air. A highly differentiated bone response develops against this oxide layer when the implant is buried in bone. This intimate bone healing around titanium implants has been called osseointegration. One study has claimed this contact should involve a minimum of 90-95% of the implant surface. As well as encouraging intimate bone healing, these titanium implants have been shown to preserve marginal bone height and reduce alveolar ridge resorption. There are of course many other types of endosseous cylinder implants available, some with very promising success rates.

Several types of hydroxylapatite coated cylindrical root form implants have recently been introduced. These include IMZ-HA (Interpore International), Integral (Calcitek), Sterioss (Denar) and Bio-vent (Core-Vent).

These implants have a range of diameters from 3.25 to 4.0 mm as well as various lengths. The narrower implants can be placed in areas of reduced bone width.

The implant core is manufactured from titanium or titanium alloy. Dense hydroxylapatite is known to be biocompatible and to act as a

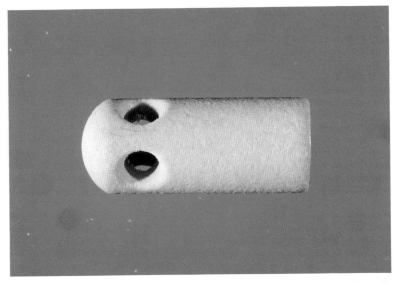

Fig. 7.10 A Calcitek Integral implant with hydroxylapatite coating applied all the way up to the interface.

scaffold for new bone formation. The hydroxylapatite coating aids direct bonding of bone to the implant surface. Bone is thought to be laid down on the coating and on the site of the surgically prepared socket. This has the effect of closing the surgical site more quickly as new bone grows from the implant surface to meet bone growing from the socket. These coated implants have also been used in conjunction with bone grafting materials, for example demineralised bone and hydroxylapatite, to enable implants to be placed where there is inadequate bone. A typical example of this has been the development of sinus lift procedures to augment the atrophic maxilla.

The use of hydroxylapatite coated root form implants
Much of the implant clinical work detailed in this book revolves around this particular design, an example of which is the Calcitek Integral implant which was developed in 1984 (fig. 7.10). It consists of a titanium alloy core which is textured to enhance the adhesion of hydroxylapatite. The core is coated, using an argon plasma spraying process, with a 50–75 micron layer of dense hydroxylapatite.

This produces a uniform, highly crystalline coating free from contamination (fig. 7.11). The implant is cylindrical in shape and has two

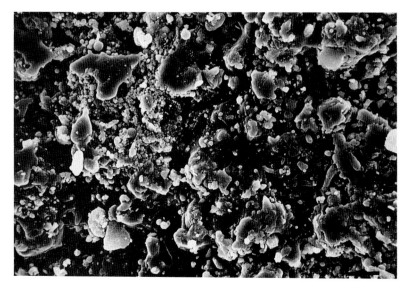

Fig. 7.11 A scanning electron micrograph of a Calcitek 4.0 mm Integral implant shows a uniform highly crystalline hydroxylapatite coating (1000 × magnification).

diameter sizes (3.25 and 4.0 mm). Four different lengths are commonly available (8, 10, 13 and 15 mm). The implants are surgically placed using a two stage procedure. The placement is comparatively simple compared to other techniques because there is no need to form an internal thread in the bony socket, as the implant is just a push fit.

A 5-year clinical follow up of the Calcitek implant has reported a success rate, in terms of functionally retained implants, in excess of 95%. This study reported on the placement of 772 implants of which only 29 had failed.

There are some disadvantages to this implant in that every effort must be made to avoid contamination of the surface when placing the implant. The hydroxylapatite coating cannot be cleaned and re-sterilised unlike titanium. Rough handling when placing the implant may cause damage or fracturing of the surface coating and infection around the implant site may lower the local pH levels enough to cause resorption of the hydroxylapatite.

Finally as the implant is cylindrical and a push fit, there is no initial stabilisation. Care has to be taken not to apply pressure from a denture which may be transmitted to the implant and cause movement in the early phase of healing.

The evidence for root form implant supported overdentures

There are as yet few long term evaluations of implant supported overdenture treatment. The majority of studies have looked at implant placement in the mandible and all report a high implant survival rate. Sixty-four edentulous patients who were having severe problems with conventional mandibular dentures were treated with implant supported overdentures. They were followed for up to 6 years with a 97% implant survival rate. Ninety-five per cent of the patients were satisfied with the overdentures provided.

Following treatment patients reported a marked improvement in self-confidence and social rehabilitation. There are few reports on maxillary overdentures but most show a lower implant survival rate (85%) which is thought to be due to the poorer quality and quantity of bone in the upper jaw. The necessary follow-up maintenance was more extensive and the gingival tissues appeared more prone to hyperplasia. The retentive clips and the denture also seem more likely to fracture.

Indications for using implants to support overdentures

The problem complete denture patient

This type of patient is a major reason for selecting dental implants as a treatment option. It is known from the most recent Adult Dental Health survey (1988) that about 10% of all complete denture wearers were not satisfied with their dentures. The majority of complaints relate to mandibular dentures which they find difficult to control and which are painful and interfere with chewing ability. These patients frequently have advanced residual ridge resorption and atrophic denture bearing mucosa which reduce the chances of successful denture wearing.

The ability to place dental implants in such conditions offers a new solution to denture stability and retention. Dental implants have been shown to improve the patients' oral function dramatically to almost the level of the natural dentition. Patients converted from mandibular complete dentures to overdentures supported by implants produced a more consistent chewing pattern, possibly due to increased stability and retention.

Poor soft tissue support and speech

Aesthetic problems can arise in the edentulous patient and are frequently associated with poor lip support. In the lower jaw this may be due to a deep labiomental groove. In the upper jaw lack of lip support combined with a severely resorbed maxillary arch is not easily treated with a fixed prosthesis supported by implants because of the lack of a denture flange. The gap between the bridge and the underlying soft tissue can allow the passage of fluid or even permit small particles of food to escape from the mouth. This passage of air and fluid could make it difficult for the patient to speak clearly. An overdenture flange would replace the missing soft tissue and avoid these problems. There is also more freedom with an overdenture to arrange the teeth and flange to give optimal lip support in both jaws.

Patients with retching problems

Implants can be a very useful aid in the treatment of patients with retching problems. A fixed prosthesis gives added security and is the first choice. If there is insufficient bone to place enough implants to support a fixed prosthesis then a horse-shoe type of prosthesis is the only design likely to be tolerated. Two or preferably four implants would be enough to support such a prosthesis.

High occlusal loading and parafunction

Where occlusal loading is likely to be high, an overdenture can have the advantage of spreading the forces over the available denture bearing tissues thereby reducing damaging forces on individual implants. In a patient with parafunctional habits a removable prosthesis may be easier to maintain and adjust as wear takes place.

Congenital or acquired structural defects

Implants supporting an overdenture can be particularly useful where there is a congenital or acquired structural defect to the jaw which will seriously affect denture stability and retention. The common examples are cleft palate, surgical defects following tumour removal, and trauma resulting in hard and soft tissue loss. The greater stability offered by an implant supported obturator can be of major benefit to patients with large oral defects (figs 7.12 and 7.13).

Alveolar ridge atrophy

Marked ridge resorption may result in the use of shorter implants which would be less able to withstand high occlusal forces transmitted via a fixed prosthesis. Poor bone quality would have a similar effect. An overdenture enables the masticatory forces to be dispersed onto the tissues as well as the implants. A resilient attachment would allow movement of the denture and some degree of stress relief.

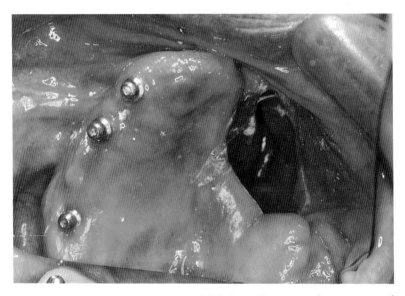

Fig. 7.12 Three Brånemark implants with ball attachments used to support and help retain a maxillary obturator.

Financial limitations
As an overdenture uses fewer implants and is easier to fabricate, it has financial advantages if cost is a problem for the patient. If at a later date the patient should wish to transfer to a fixed prosthesis, and bone support allows this, further implants can be placed.

Reduced number of implants
Finally, overdentures offer a solution to the restoration of those patients in whom for some reason there has been a reduction in the number of planned implants so that a fixed prosthesis is not possible. In a similar vein it may be necessary to convert a patient from a fixed prosthesis to an overdenture if they are unable to maintain their plaque control or have some other reason for not accepting their fixed prosthesis, such as poor lip support due to the lack of a labial flange.

The type and number of implants used in each patient is dependent on the bone quality and quantity and the aesthetic and economic demands of the patient.

Contraindications to implant retained overdentures
There are few contraindications to implant supported overdentures. The main ones are gross alveolar ridge atrophy or bone of such poor

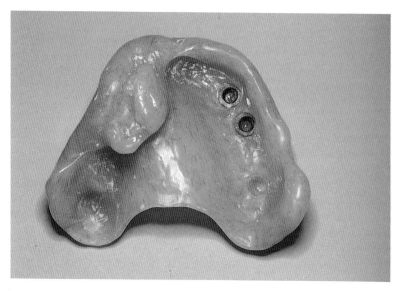

Fig. 7.13 Fitting surface of obturator showing plastic caps and O-rings in position.

quality that it would have a very adverse effect on implant integration. Patients unable to accept a removable prosthesis should not be considered.

Patient selection

It is obvious that not every patient is suitable for the placement of dental implants and indeed not every patient would wish to have them. The patient has to be carefully evaluated and alternatives to implants discussed with them. The selection process is probably the most important aspect of dental implant treatment. Patient selection is determined by the clinicians involved. Generally this involves an oral surgery specialist and a prosthetist. The availability of articulated casts, appropriate radiographs, and diagnostic wax-ups are helpful when explaining treatment to the patient.

Medical considerations

There are some medical contra-indications to dental implant treatment and in particular those conditions which compromise healing response such as uncontrolled diabetes, long-term steroid therapy, bleeding disorders, high dose irradiated patients, immunologically compromised patients, severely debilitated patients and those with heart defects which may predispose the patient to subacute bacterial endocarditis. In addition it would be well to avoid patients with potential drug problems, high alcohol intake, heavy smokers and those with known severe psychiatric disorders as they are unlikely to be able to maintain an adequate level of oral hygiene.

Psychological assessment
Generally there is no need to evaluate all patients psychologically and there is evidence to suggest that clinicians are not very accurate at assessing personality traits. Some clinicians may find a simple self assessment personality questionnaire useful. It is probably more important to recognise patients who have unrealistic expectations with regard to the final result. Patients who are taking psychotropic medication may present more of a risk mainly in terms of poor compliance throughout treatment and subsequent maintenance. The overall view is that many of these patients can be successfully treated if they are given psychological support throughout their treatment. The more severe the psychiatric condition the greater the risk of failure and these patients should therefore not be considered suitable for implant retained prostheses.

Anatomical assessment
A thorough clinical examination, including study casts and radiographs, is essential. Associated structures need to be checked, especially where there is marked ridge atrophy. The position of the mental foramina and inferior dental canals should be identified in the mandible. Implant length in the maxilla is limited by the position of the nasal cavities, sinuses and incisive canal.

The patient must have adequate bone height and width to accept the selected implant. There should be at least 8 mm of usable bone height and 5 mm of bone width. There must be enough bone available to place at least two and preferably up to four implants. The implant site and surrounding areas should be free of local pathology. Most of this can be confirmed by taking appropriate radiographs. The overlying soft tissues should be healthy and there should be a reasonable amount of attached keratinised gingival tissue. Finally there must be enough room for the planned restoration.

Denture wearing assessment, motivation and hygiene
Past denture history enables an accurate assessment of the patients needs. It is important to examine existing dentures with regard to tooth positioning, lip support and occlusal vertical dimension since these

features may require alteration in order to make enough room to place implants and their attachments. If this is done prior to implant placement the patient is able to visualise the changes. Finally there must be a genuine reason why a conventional approach to treatment cannot be used. All reasonable efforts should be made to correct underlying prosthetic errors before implants are given serious consideration. The patient should be well motivated and capable of good oral hygiene in order to maintain the implant and attachments.

Informed consent

It is obviously important that all the members of the team and the patient agree on the final treatment. The patient needs to understand the stages involved and the likely outcome of the treatment. Risks and possible complications should be fully discussed.

8

Implant Retained Overdentures: Clinical and Technical Procedures

This chapter describes the planning required prior to implant placement. It also details the surgical placement of a two stage root form implant. Finally, four types of overdenture abutment attachment are discussed.

The team approach

The success of dental implants depends greatly on a team approach, with each team member understanding their own particular role. The full team will often consist of specialists in oral surgery and restorative dentistry, a dental technician and a hygienist. All members of the team should be familiar with the implant system they are using and must have undertaken recognised postgraduate training.

It is quite feasible, especially in overdenture cases, for one suitably trained clinician to both place the implants and look after the prosthetic aspects. Liaising with an oral surgery specialist can help with the treatment of more difficult cases, particularly those which may involve bone grafting or sinus lift procedures.

Whatever the make up of the team it should be the restorative dentist who initiates treatment and plans the final prosthetic restoration. It is to the restorative dentist that the patient will return for future clinical care and maintenance. The clinician must understand fully the treatment plan and be able to design the prosthesis that will eventually restore the patient's mouth.

The dental technician has a vital role in ensuring that the final prosthesis is fabricated to the highest standards. There must be good communication between technician and clinician to ensure that the former has all the information available to follow the treatment plan. The clinician must provide accurate impressions and records of the occlusion for evaluation. They must understand the limitations of proposed technical procedures and should be familiar with any attachments that are used. Joint planning sessions are particularly valuable when difficulties are expected due to the chosen implant site or angulation of abutments.

The oral surgeon takes responsibility for the placement of the implants and needs to be able to assess accurately the quality and quantity of bone available. Of equal importance is the need to know the prosthetic limitations of the system that is being used. Occasionally a situation occurs where there appears to be enough bone to place implants but the path of insertion would result in a poor emergence angle. Such a positional problem would pose technical difficulties, compromise aesthetics and lead to unfavourable loading of the implant.

The hygienist plays a very important role in the maintenance of the implant and the surrounding gingival tissues. Many people become edentulous because of poor oral hygiene and will need educating in the care of the implant, its attachment and the final denture. The implants and their attachments can be easily damaged by over zealous cleaning, particularly if abrasives are used. The patient's oral hygiene will need to be reviewed at least once every 6 months in the early stages of treatment.

Pre-implant prosthetics

All patients who seek implant treatment should have their existing dentures evaluated. Implants are not a solution to poorly made dentures and many patients benefit from new dentures which correct basic errors. During a remake of dentures teeth can be moved into more acceptable positions to reduce aesthetic problems that may be posed by the emergence position of the implants. Dentures can be thickened over areas where attachments may be placed. The vertical height of lower dentures can be increased, if freeway space allows, to facilitate the later placement of attachments in the denture. The effect of these changes on speech and appearance can then be checked. At the same time radiographic and surgical templates can be constructed.

Radiographic views and radiographic templates

When placing implants there are anatomical structures which must be located and avoided such as the maxillary sinuses, nasal cavities and in the posterior aspect of the mandible, the inferior dental canals.

The radiographic examination will include extra-oral and intra-oral films in order to determine the quality and quantity of bone. When assessing bone quality it is important to look at the ratio of cortical to cancellous bone. A very high cortical to cancellous ratio is usually found in the anterior part of the mandible. A thick cortical plate surrounding cancellous bone is found in the posterior region of the mandible. In the maxilla a thin cortical plate surrounds a large amount of cancellous bone. In a patient suffering from osteoporosis there will be an increased porosity and thinning of the cortex which surrounds a low density spongy bone. The prognosis for placing implants in such a patient will be poor.

The shape of the jaws needs to be assessed. A ridge with minimal bone resorption presents few problems. A ridge with progressively severe resorption into the basal bone may require bone graft augmentation before implants can be placed.

The minimum bone needed for Integral implant placement is 5 mm of ridge width and 8 mm of bone height. It is always worth selecting the maximum implant length possible to give the best chance of withstanding occlusal loading.

A panoramic radiograph is an excellent general view for evaluating available bone and identifying anatomical and pathological conditions. The magnification of these radiographs can be calculated by taking the views with a radiographic template in place. Several designs exist and an example of one type is shown in figure 8.1. This consists of a well fitting autopolymerising base constructed with vertical and horizontal metal wires of known lengths. As well as measuring bone height the rods can be used as reference points to locate structures in the vicinity of possible implant sites.

The superimposition of the cervical spine on orthopantomograms may result in some loss of definition in the anterior region. A lateral cephalogram is a useful view to determine correct bone height, shape, width and angulation of the premaxilla and mandibular symphysis. It only provides an image of the mid-line and does not reflect the anatomical outline of the lateral segments of the jaw. It can be supplemented by tomograms cut at planned intervals from the mid-line backwards.

A true occlusal film will show the edentulous jaw width as a plan view and may be a useful additional radiograph. If implants are to be placed in the posterior parts of the mandible or maxilla then computerised tomography (CT) scanning may be useful in preoperative evaluation. A CT scan can give sectional views of the jaw using cuts of about 1.5 mm. This type of scan can now give three-dimensional views of the jaw to be operated on and is so accurate that subperiosteal implants have been manufactured on the models produced from such an evaluation. The main disadvantage of CT examination is the much higher radiation

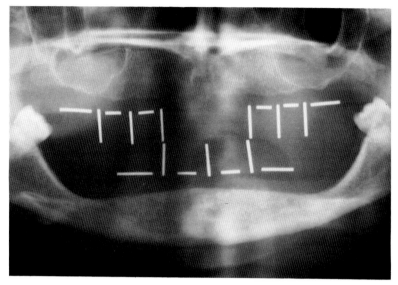

Fig. 8.1 A diagnostic radiograph taken with a template in place. The wires are of known length and aid in selecting final implant size and positioning.

exposure (5-10 times higher than conventional tomography) to the patient. If metal restorations are present then the radiation is deflected into the surrounding tissues. Finally, the equipment is not readily available and is expensive to purchase and use.

Implant placement and surgical templates

The placement of the Integral implant is described here but similar methods govern the use of most root form implants. As with any surgery performed in the mouth it is vital that aseptic techniques are followed. The surgery itself is carried out in two stages. Firstly the implants are placed and then after 3 months in the mandible and 6 months in the maxilla they are uncovered. Both stages can usually be completed under local anaesthetic. Occasionally relative analgesia or sedation will be required.

The implant bodies and healing caps are provided sterile, ready for implantation. Due to the nature of the hydroxylapatite coating the implant body cannot be re-sterilised if it is contaminated prior to placement.

If new dentures have been constructed prior to implant placement they can be duplicated in clear acrylic resin. These duplicates can then be used to fabricate surgical templates, of which there are several

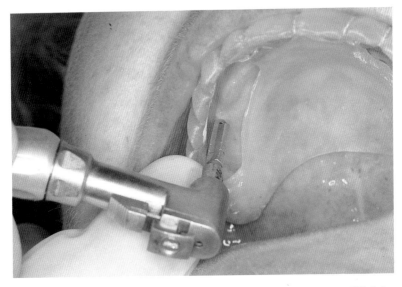

Fig. 8.2 A clear resin duplicate of the patient's maxillary denture modified for use as a surgical template. A slot has been prepared to act as a guide during implant placement.

designs. The idea of a template is to guide the surgeon when placing implants. The implant should emerge lingually/palatally in the region of the cingulum of the anterior teeth and through the central fossa of the posterior teeth.

A template which can be used in a situation where there may be some doubt about the amount of bone available for implant placement is shown in figure 8.2. In this situation all the material lingual to the teeth is removed leaving a thin labial veneer. This veneer is cut down to allow access for the contra angled handpiece. The template is also relieved from the fitting surface to allow space for the mucoperiosteal flap. This template leaves the surgeon more room to select the optimum placement for the implant whilst still having the ability to check on the labial placement of the teeth.

Implant placement involves a mesiodistal incision along the buccal aspect of the alveolar crest. The incision should be long enough to permit adequate flap resection and, if required, a vertical incision should be used. A periosteal elevator is used to lift the mucoperiosteum and expose the alveolar bone. Flap retractors are used and bone irregularities are smoothed using a rosette bur or rongeurs forceps to create a flat bone surface at least 5-6 mm wide (fig. 8.3). Following any ridge reduction it

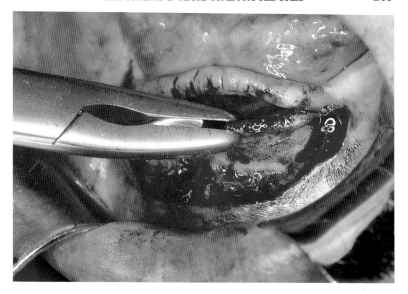

Fig. 8.3 The removal of sharp bony ridges to create a flat surface at least 5-6 mm wide.

is a good idea to confirm the selected implant size. The internal drilling of the bone should be completed at slow speeds (less than 850 rpm) using a high-torque, internally irrigated handpiece. The irrigation will minimise excessive heat generation and preserve the vitality of the surrounding bone.

The range of drills used is shown in figure 8.4. The cortical plate of the ridge crest is penetrated with the pilot drill. This hole will establish the angulation of the implant. It should be prepared using the surgical template as a guide. On completion of the first hole the bone debris is flushed out and one of the parallel pins is inserted using the end which corresponds in size to the pilot drill. This pin, alongside the surgical template, will help to guide the angulation used on the next hole preparation. After each hole is prepared further parallel pins are used (fig. 8.5). When all the pilot holes are completed the parallel pins are checked for correct angulation and spacing. The parallel pins are removed and a rosette bur is used to create a dimple on the surface of the pilot hole to aid the start of the intermediate spade drill. The intermediate spade drill is marked with concentric rings corresponding to various implant lengths of 8, 10 and 13 mm. The rings are 1 mm longer than the corresponding implants to compensate for irregularities in the

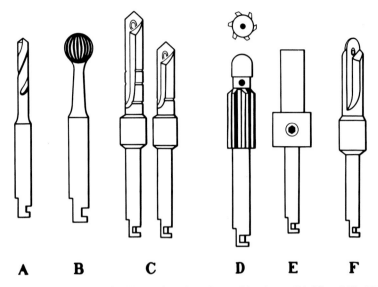

A   B   C   D   E   F

Fig. 8.4 The selection of drills used to place Integral implants. (A) Pilot drill, (B) Rosette bur, (C) Intermediate drill (note the concentric rings which correspond to various implant lengths, (D) Countersink drill, (E) Drill extension, (F) Final drill.

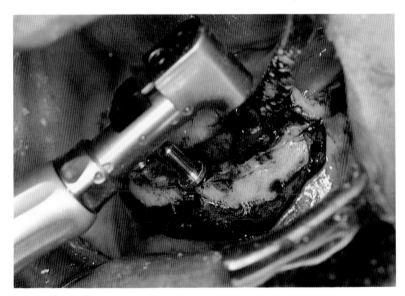

Fig. 8.5 Initial pilot hole preparation using parallel pins as a guide to the path of insertion.

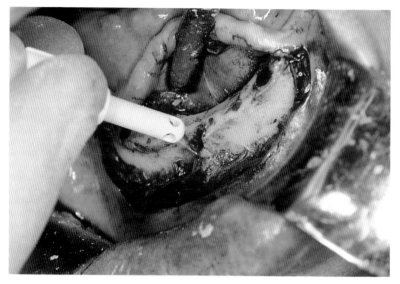

Fig. 8.6 A Calcitek implant is placed using the plastic cap as a handle to avoid contacting the implant surface.

alveolar crest. Small corrections to aid parallelism can be made at this stage. Once again alignment can be checked using the other end of the parallel pin. After the intermediate spade drill, a countersink drill is used to widen the cortical bone to the same diameter as the final spade drill. Finally, the full diameter spade drill is used. Each drill has a diameter and length that corresponds to a specific Integral implant body.

The prepared site is irrigated and the implant is positioned by holding the plastic cap (fig. 8.6). The implant is seated as far as possible and the cap is then snapped off. The titanium healing screw remains screwed into the implant body but should be checked for tightness of fit. The implant can then be fully seated using a gentle tapping action with a rubber mallet onto a plastic tipped tapper. The implant should sit flush with or just below the crest of the alveolar bone. The wound is then closed and the patient advised not to wear the denture for a period of 2 weeks. It is important that the implants should remain unloaded during the early phase of healing. Immediately after surgery, analgesics and antibiotics of choice may be prescribed. After 2 weeks the patient's existing denture should be modified to remove any pressure applied directly to the buried implant. The denture can be relined at this stage

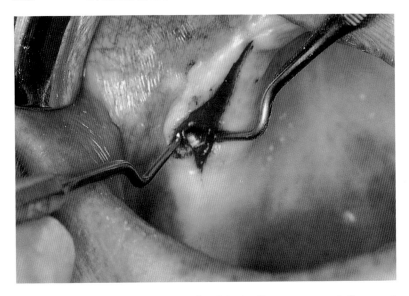

Fig. 8.7 The implant has been exposed and the healing cap removed. Care must be taken to remove all bone and soft tissue from the superior aspect of the implant.

with a short-term resilient lining material. The patient will need to be seen occasionally during this healing phase to monitor healing and to adjust the fit of the denture.

Surgical exposure of implants

Following a healing period of 3 months in the mandible and 6 months in the maxilla the implants can be uncovered. The location can usually be determined by palpating the soft tissue. The implant can be exposed, either by using a tissue punch or a scalpel. The objective is always to try and preserve the attached gingival tissue. Before attempting to remove the healing cap all bone and soft tissue must be cleared from the superior surface of the implant (fig. 8.7). This is to insure that the transmucosal abutment will seat completely onto the surface of the implant. Great care is needed at this stage not to damage the implant body. A temporary gingival healing cuff is screwed into the implant body (fig. 8.8) and the patient's existing denture is eased over the cuffs. It is important to avoid trauma to the healing gingival tissues. Retention and stability of the denture may be improved by a further addition of short term resilient lining material (fig. 8.9). Prosthetic restoration can begin once soft tissue healing has taken place, usually within 2-3 weeks (fig. 8.10).

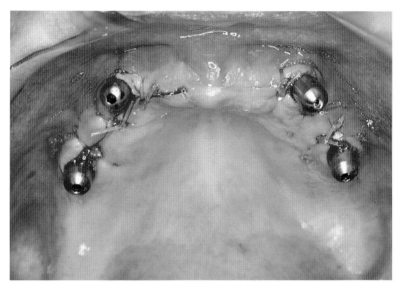

Fig. 8.8 The exposed implants with temporary gingival healing cuffs in place.

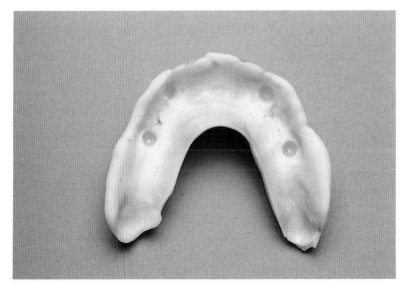

Fig. 8.9 Patient's denture modified with the addition of a short term resilient lining material to fit over the temporary gingival healing cuffs.

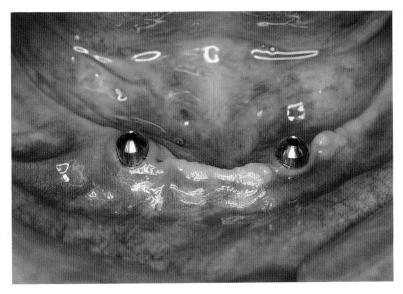

Fig. 8.10 Three weeks after exposure the gingival tissues in the lower jaw have healed sufficiently to record working impressions.

Types of implant attachment

There are many types of precision attachments that can be used. In fact, virtually all the attachments described earlier in this text relating to conventional overdentures can be used with implants. In this chapter four simple but effective attachments that have been consistently used with the Integral system will be described.

The O-ring attachment

The Calcitek O-Ring attachment consists of three parts, the titanium abutment, the metal housing and a black rubber O-Ring (fig. 8.11). The titanium abutment is designed as a small ball with a retentive undercut groove. This is screwed into the implant using a special seating tool which has an internally machined hexagon designed to engage the abutment's external hexagon. The seating tool holds the abutment securely while it is being placed into the patient's mouth. The abutment's gingival cuff ranges from 2 to 5 mm in length. The cuff size selected depends on the depth of the pocket around the implant. The attachment requires at least 7 mm of height coronal to the implant to allow for the abutment and the retaining ring mounted in the denture. The abutments must be more than 7 mm apart to allow room for the metal housing. They should not converge or diverge

Fig. 8.11 The O-Ring components consist of the titanium abutment and the O-Ring metal housing which contains the black rubber O-Ring.

by more than 10 degrees, or difficulty will be encountered in seating and removing the overdenture with accelerated wear of the black rubber O-Rings. A metal housing retains the black O-Ring and is processed into the denture. The black rubber O-Ring engages the undercut groove on the abutment to provide resilient retention.

Clinical technique
If the patient has a satisfactory denture the O-Ring metal housing can be fitted to the denture in the mouth. The temporary gingival healing cuff is removed and the soft tissue pocket depth is measured with a periodontal probe. The appropriate O-Ring abutment is selected so that the cuff is at least 1 mm clear of the gingival margins (fig. 8.12). The impression surface of the denture is relieved in the area directly over the abutment. A hole of 6 mm in diameter is prepared and is vented lingually to allow excess acrylic resin to escape. The black O-Ring is fitted to the metal housing and is seated onto the abutment. The denture is repositioned to verify that adequate relief has been allowed around each abutment. After this the bottom of the metal housing down to the soft tissue is blocked out with wax to prevent excess acrylic resin from locking the denture onto the abutment (fig. 8.13).

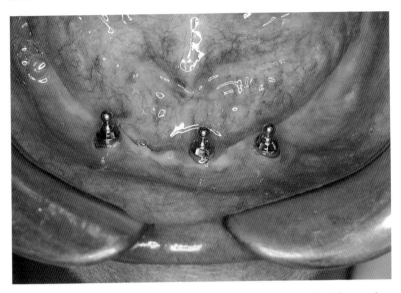

Fig. 8.12 The O-Ring titanium abutments seated with their cuffs 1-2 mm clear of the gingival tissues.

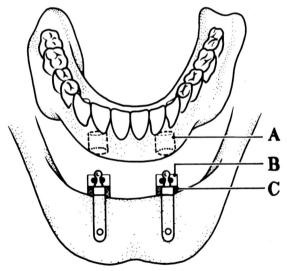

Fig. 8.13 Modifying an existing denture to accept O-Ring attachments. (A) The denture is relieved directly over the attachments. (B) The black O-Ring and housing is fitted to each attachment. (C) The space between the housing and soft tissue is blocked out.

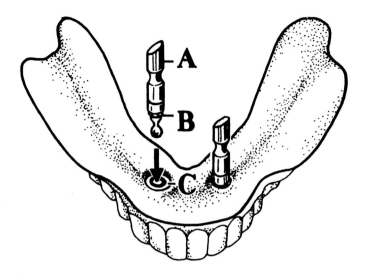

Fig. 8.14 Laboratory addition of O-Ring attachments. A = implant analogue, B = O-Ring abutment and C = reline impression of O-Ring abutment taken *in situ*.

Autopolymerising acrylic resin is then applied to the top of the housing and to the prepared holes in the denture; the denture is seated and brought into occlusal contact. After the acrylic resin has cured, the black O-Rings are removed from the housing and a small round bur is used to relieve the acrylic resin above the spherical post to prevent the denture base from contacting the top of the post when seating. Contact between the top of the abutment and the denture base can cause loosening of the abutment and reduced resilience of the overdenture. When multiple abutments are used it is recommended that only two are processed at each time.

The addition of the O-Rings can be undertaken in the laboratory by first relieving the denture as described above. A reline impression is taken in the denture of the abutments without the O-Ring or housing in place. The denture and impression are removed and the abutments unscrewed from the implants. The implant analogue is screwed into the O-Ring abutment and the assembly is replaced into the impression (fig. 8.14). A model is cast and the O-Ring and housing can then be processed into the denture in the normal way.

When constructing a new denture an alternative method is to take working impressions of the denture bearing area and of impression posts which have been screwed directly into the implant (fig. 8.15). The

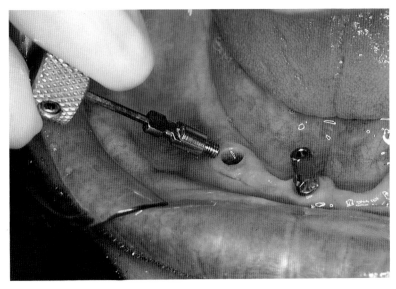

Fig. 8.15 Impression posts being placed directly onto the implants using a hex driver.

impression is removed and the impression posts unscrewed from the implants. The posts are now attached to the implant body analogues and reseated back into the impression (fig. 8.16). A model is then cast and the impression posts removed. The O-Ring abutments can now be screwed into the model. The housings can then be processed into the final denture (fig. 8.17).

The O-Ring supported prosthesis is a very simple but effective form of overdenture. It facilitates good oral hygiene and maintenance. The black O-Ring is economical and easy to replace should signs of wear become noticeable. It does not take up a great deal of space and is suitable for those cases with limited inter occlusal clearance.

Ball overdenture attachment
This is a very simple but relatively bulky overdenture attachment. It consists of a spherical head and shaft with a variable cuff length (fig. 8.18). The cuff length varies from 2 to 5mm depending on pocket depth measurements. The main contra-indication for this attachment is insufficient space available to accommodate it and the soft lining material used for retention within the denture. It is also unsuitable when divergence between two implants is greater than 15 degrees.

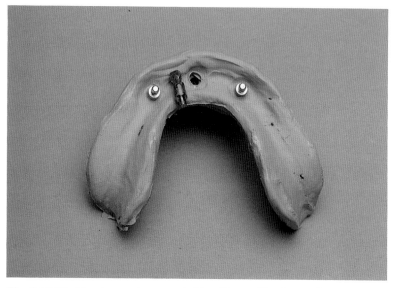

Fig. 8.16 Working impression recorded in silicone. The impression posts must fit precisely back into the impression.

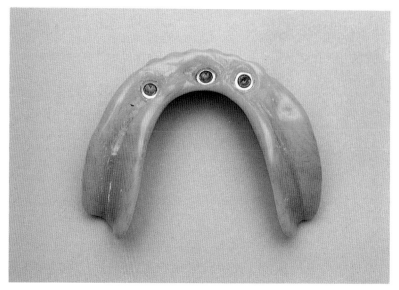

Fig. 8.17 Impression surface of O-Ring denture with black O-Rings in place.

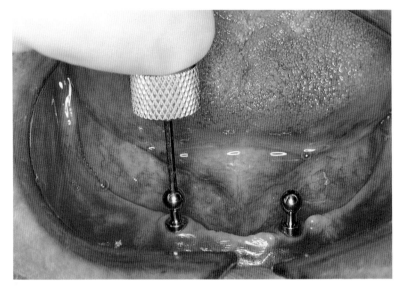

Fig. 8.18 Ball overdenture attachments secured to the implants.

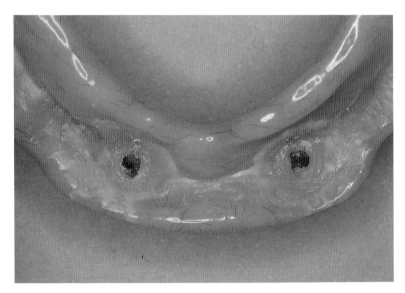

Fig. 8.19 Impression surface of the denture with heat cured soft lining material processed around the ball overdenture abutments.

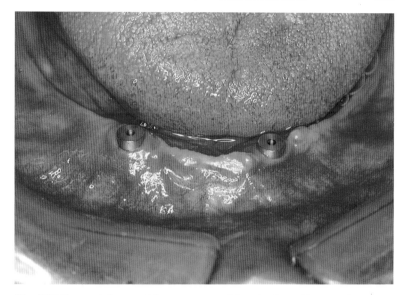

Fig. 8.20 Two stainless steel keepers screwed into implants. Rare earth magnets will be used to provide retention.

Clinical technique
A minimum of two implants must be used which should usually be sited bilaterally in the canine region. The temporary gingival cuffs are removed and replaced with attachments of appropriate lengths. An impression is then taken and when removed transfer analogues of the ball attachments are seated back into the impression. A model is cast and retention is achieved by processing a conventional heat cured silicone soft lining material around the portion of the sphere that protrudes supragingivally (fig. 8.19).

Magnetic attachments
These consist of three components namely: (a) a magnet, (b) a stainless steel keeper, (c) a magnetic attachment cuff. There are multiple lengths for both the keeper and the cuff (fig. 8.20). There are still some doubts about the use of magnets as heat cured processing is thought to reduce the magnetic force. There is concern about corrosion of magnets in oral fluids and the effect that corrosion by-products might have on the critical tissue implant interface. Their clinical use in implant retained overdentures is decreasing and some manufactures no longer stock components. A magnet retained overdenture should not be used in a very resorbed ridge as it does not provide good lateral stability.

Clinical technique

A minimum of two implants must be used. The temporary gingival healing cuffs are removed. A magnetic cuff and stainless steel keeper are screwed firmly into the implant body leaving the keeper slightly supragingival. Black impression pieces are placed into the keepers (fig. 8.21). An impression is now recorded capturing the black impression pieces. White keeper analogues are fitted to each black impression piece to replicate the intra-oral keepers. A master cast is then poured. On this cast a black processing piece is located onto each keeper analogue. The final denture is processed with the processing pieces retained in the denture base. The processing piece is unthreaded from the denture base and is replaced with the magnet. The magnet can be adjusted to maintain close contact with the keeper (fig. 8.22).

A bar retained overdenture

There are a number of bar attachments that can be used with implants. The implants need to be placed so that they are in the centre of the ridge and the bar can follow the ridge crest. The denture is usually secured by means of gold or plastic clips. The main advantage of a bar system is that it provides excellent retention and splints the implants together. The main disadvantages are the greater risk to oral hygiene and the difficulties encountered when attempting to reline the denture or repair the clips. The bar also takes up space within the denture and can lead to potential weakness and denture fracture.

A popular plastic bar attachment is the 'Hader bar'. The bar is available in a prefabricated plastic form which should not be bent or curved. In cross-section, the bar is round-shaped in the occlusal part and flat in the lower portion. A 2.0 mm gingival clearance between the bar and gingival tissues is necessary for good oral hygiene. The bar is a straight connection between the abutments. The prosthesis can then rotate slightly in an antero-posterior direction to allow for tissue resilience. If the bar is allowed to curve the plastic clip would be very tight over the bar. This would transfer more force to the two implants.

Clinical technique

The temporary gingival healing cuff is removed and the correct length of shouldered abutment fitted. The cuff should be clear of the gingival margins by 1-2 mm (fig. 8.23). Impression transfer copings are screwed onto the shouldered abutment and impressions taken. The impression is removed and the transfer coping unscrewed from the shouldered abutment. The transfer abutment is then screwed onto an abutment

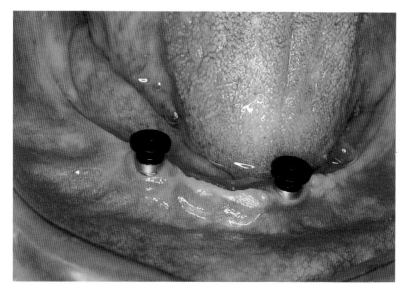

Fig. 8.21 Black impression transfer pieces placed onto the steel keepers prior to recording the working impression.

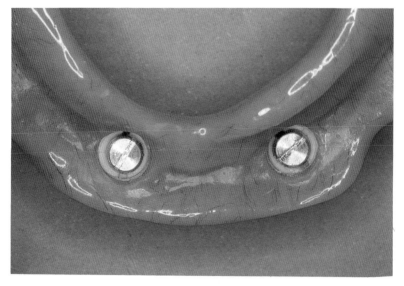

Fig. 8.22 Impression surface of denture with magnets in place.

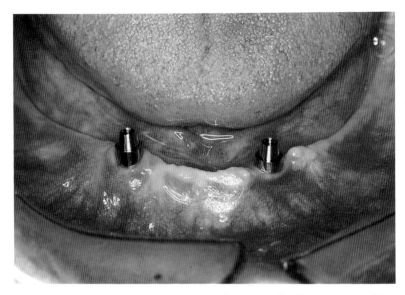

Fig. 8.23 Shouldered abutments screwed down onto implants prior to taking working impression for Hader bar. The shoulder should be 1 mm clear of gingival margin.

analogue and inserted back into the impression. A cast is poured and the transfer coping removed. Waxing sleeves are now secured to the abutments with a coping screw. Waxing sleeves can be adjusted to a level suitable for the bar attachment. The plastic Hader bar is trimmed to allow a 2 mm space for cleaning and attached to the waxing sleeves with wax (fig. 8.24). A silicone jig containing the denture teeth can then be tried in over the bar. The position of the bar in relation to the teeth can be checked (fig. 8.25). The bar is then sprued and cast in gold alloy. The cast bar is placed back on the master cast. The plastic clip used in processing the final denture is fitted to the bar. The undercuts are blocked out and the model duplicated if required prior to processing. When the denture has been processed the plastic spacer is removed. The final clip is then placed into the overdenture using the clip holder (fig. 8.26).

The bar attachments take up considerably more room within the denture than the other attachments mentioned (fig. 8.27). This can make the denture prone to fracture and it should therefore be processed in high impact acrylic resin.

There are numerous other attachments that could be used in conjunction with implants. Most clinicians tend to restrict themselves to

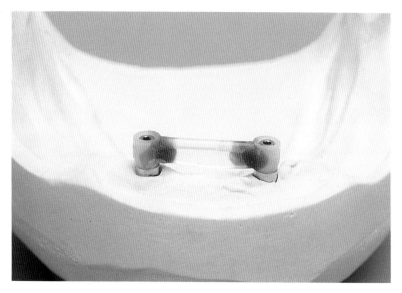

Fig. 8.24 Waxing sleeves secured to abutment analogues have been adjusted for height. The plastic Hader bar has been attached to waxing sleeves with 2 mm ginigival clearance.

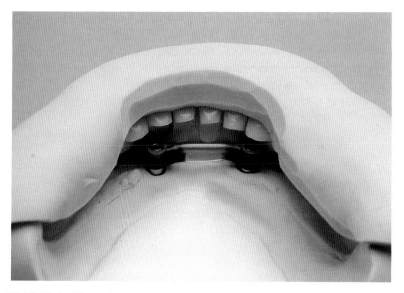

Fig. 8.25 A silicone jig containing the denture teeth is tried over the waxed-up Hader bar to assess the room available for this attachment.

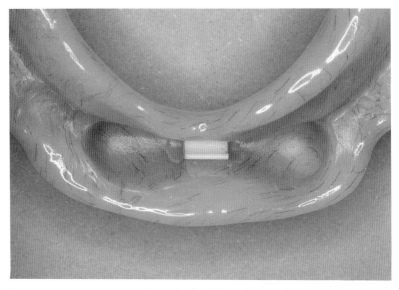

Fig. 8.26 The plastic retentive clip fitted into the overdenture.

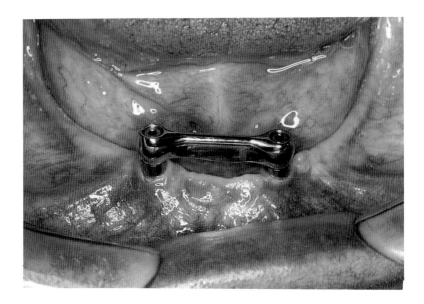

Fig. 8.27 Try in of cast bar secured onto implants.

two or three varieties which deal with the majority of clinical situations. Simplicity, ease of maintenance efficiency and economy are the usual selection criteria.

Impression procedures are very similar to those already described in Chapter 5. Particular care is needed when the denture bearing tissue of the saddle area distal to the implants is compressible. Unless a mucocompressive impression technique is used the denture may pivot around the implants and apply high horizontal torque loads to them. Excessive loading will lead to bone loss around the implant and its premature failure. The authors routinely green stick the saddle areas and relieve the tray over the implants in order to achieve tissue compression. Accurate impressions are essential to the success of implant retained overdentures.

9

Implant Retained Overdentures: Complications and Maintenance

This final chapter discusses the common complications and maintenance requirements of an implant retained overdenture. With good planning, adequate clinical reviews and excellent home care complications are few and maintenance is minimal.

Complications

There are various complications that can arise with implant treatment. After the first stage of implant placement the patient is advised not to wear the denture for at least 2 weeks. Following this it is important that the dentures have been adequately relieved over the site of the implants. A poorly adjusted denture may lead to dehiscence and exposure of the implant (fig. 9.1). If exposure occurs within the first 4 weeks of placement and plaque control is poor, the problem should be treated by raising a flap and closing over the implant. If exposure occurs much later it can be left until the second surgical stage is completed.

When the implant is exposed the healing response of the bone implant interface is checked. If movement of the implant is detected at this stage then the implant has failed and should be removed. In the absence of local infection it should be possible to consider placing a new implant in or near this site after a 3-month healing period.

At exposure there can be problems with abutment connections. Failure to seat the abutment can result in sinus formation (fig. 9.2). It is always advisable to take radiographs following the placement of the second stage abutments to check that they have seated correctly on the implant surface (fig. 9.3). Failure to check that adequate seating has occurred will result in a poorly fitting prosthesis as well as gingival problems.

A form of proliferative gingivitis can develop around an implant. This commonly occurs where there is a deep pocket (4 mm or more). It

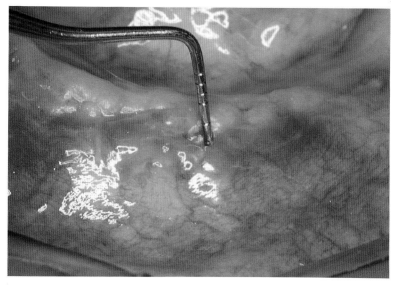

Fig. 9.1 Premature exposure of an implant following trauma from a denture.

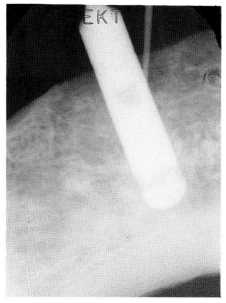

Fig. 9.2 Radiograph showing a sinus tracking from abutment implant interface caused by failure to completely seat the abutment.

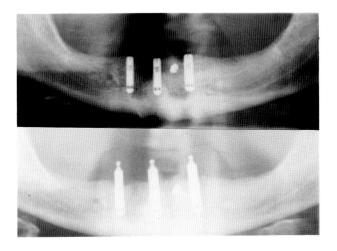

Fig. 9.3 Radiograph taken before and after seating of O-Ring abutments onto implant surface. This checks abutment seating and bone levels around implant soon after exposure.

may be due to a loose abutment or, as in fig. 9.4, where remnants of sutures placed at exposure have led to the formation of granulation tissue.

In some patients with superficial mentalis muscle attachments there may be a tendency for the tissues to grow over an exposed implant placed in the anterior part of the mandible (fig. 9.5). The patient may find it extremely difficult to clean this area because of the tension in the lower lip. To prevent potential problems it may be necessary to consider a combination of a vestibuloplasty and a free gingival graft procedure to deepen the labial sulcus.

Due to the vagaries of available bone in a grossly resorbed ridge there is a risk that this will result in the occasional placement of divergent implants. This risk can be reduced by the use of surgical templates. Although most attachments allow for some divergence, if this is greater than 15 degrees it is likely to restrict the type of overdenture which can be provided. An example of such a problem is shown in figure 9.6. Four implants were placed in the maxilla but due to ridge atrophy the two anterior implants were placed at an angle to the posterior implants. The position of the anterior abutments, if used to support the final prosthesis, would have compromised aesthetics. As it was, the divergence

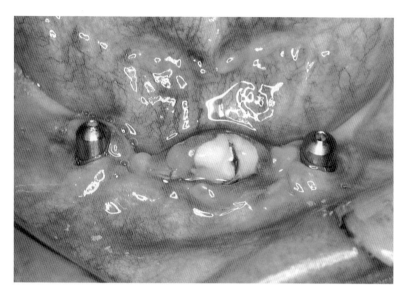

Fig. 9.4 Hyperplastic tissue forming around recently exposed implant. Thought to be caused by suture irritation.

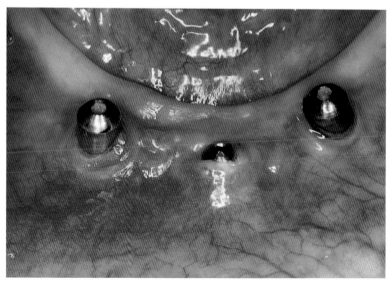

Fig. 9.5 Shallow labial sulcus combined with high mentalis muscle attachment has resulted in a partial recovering of the central implant.

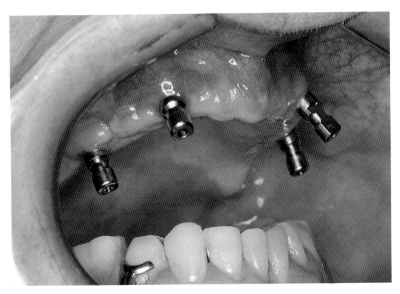

Fig. 9.6 Divergent implants can result in problems with aesthetics and function of attachments. This should not occur if a surgical template is used.

Fig. 9.7 Radiographic evidence of marked cervical bone loss around a mandibular implant retaining an overdenture.

between the implants meant that the only two suitable for retention were the posterior implants. The anterior abutments were fitted with titanium gingival healing cuffs which were left 1-2 mm proud of the surrounding tissue and thus served to support the prosthesis.

At least once a year routine periapical radiographs should be taken to assess the bone response to the implant. If progressive marginal bone loss is seen this may indicate local inflammation, occlusal overload, poor fit of prosthesis or a loosened abutment (fig. 9.7).

If the implant appears mobile the tightness of the transmucosal abutment should be checked in the first instance. If the implant remains mobile then it can be assumed that fibrous connective tissue has formed around the implant and it is best removed. There are a number of possible causes of implant failure including: (a) thermal damage to bone during implant placement; (b) the placement of an implant into an area of local pathology or contamination of the surface coating at time of placement; (c) premature exposure or trauma to the implant and surrounding tissues during the healing phase; (d) subjecting the implant to high occlusal forces or torque from attachments during the early loading phase; (e) the presence of chronic local inflammation.

The ability to move back to an overdenture from a fixed prosthesis can be an advantage in some patients. The patient in figure 9.8 was unable to maintain a high standard of oral hygiene with his mandibular bridge; converting the prosthesis to an O-Ring retained overdenture resolved this problem.

Implant maintenance

Although there is still some debate about the effects of plaque on implant health, some evidence is beginning to show that high plaque scores may lead to marginal bone loss. It is important to impress on the patient that care of the implants starts as soon as they are exposed. This can be initiated with cotton wool rolls until the surrounding gingivae become firm enough to tolerate a small soft tooth brush. An interproximal brush is ideal for cleaning around the attachments located in the denture (figs 9.9 and 9.10). During the first 2 weeks following implant exposure, a chlorhexidine mouthwash may help to reduce inflammation in the surrounding gingivae. This should not be continued when the final prosthesis is fitted as it will lead to staining.

Recall

After delivery of the final prosthesis the patient should have radiographs taken of the implant and abutments. This film, as mentioned earlier, will

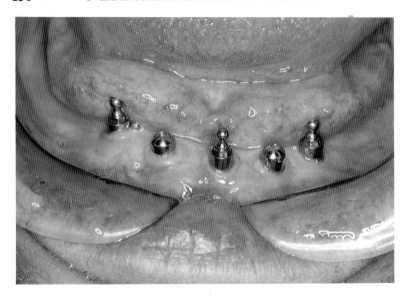

Fig. 9.8 These five implants previously supported a fixed prosthesis but on the patient's request this was converted to an overdenture retained by three O-Ring attachments.

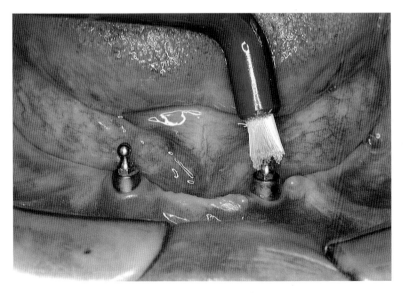

Fig. 9.9 Interproximal brush cleaning O-Ring abutments.

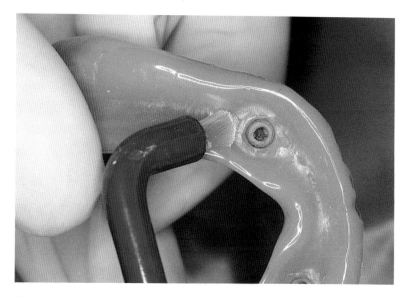

Fig. 9.10 Interproximal brush cleaning O-Ring housing.

confirm the fit of the abutment against the implant. It will also enable the bone response around the implant to be assessed. Radiographic film holders aid positioning of the film so that it is parallel to the implant. These holders can be located on templates specifically made for each patient. All subsequent radiographs can then be taken with this film holder and direct comparisons between films made.

It will be important to see the patient after one week to check the state of the implants and their attachments. Oral hygiene should be monitored and the implant abutments checked for tightness. The occlusion and the stability of the overdenture should be carefully assessed. Further reviews should be made at 1 month, 3 months, 6 months and on an annual basis thereafter. With later recalls it will be important to check for resorption along the free end saddles because marked movement of the prosthesis, particularly in the mandible, could lead to highly damaging loads being applied to the implants. A simple reline procedure will resolve this problem. The titanium surface of the abutment is easily scratched and deposits of calculus can prove difficult to remove with plastic scaling instruments. In some instances it may be easier to remove the abutment and carefully clean it out of the mouth (fig. 9.11).

With use, the attachments will begin to show signs of wear and must be looked at carefully. The rate of wear will depend on the individual

Fig. 9.11 Calculus attached to O-Ring attachment after 3 months in the mouth.

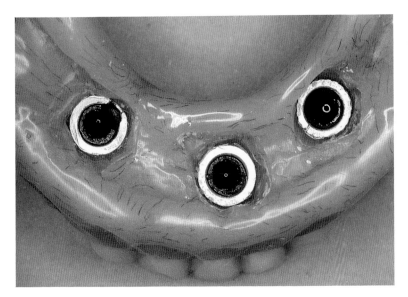

Fig. 9.12 Damage to black rubber O-Rings after 6 months of use.

patient and the divergence of the implants; the greater the divergence the higher the rate of wear. The O-Ring attachments shown in figure 9.12 have been in place for 6 months. The rubber O-Rings are beginning to collapse and should be renewed. If the metal retainer shows signs of corrosion it should be replaced.

Conclusion

The development of a simple, safe and effective way of replacing the lost natural dentition has much to recommend it. The success of modern dental implants has done much to help those patients who for a variety of reasons are unable to wear conventional dentures efficiently and comfortably. In the early years most patients seeking implant treatment were treated with a fixed prosthesis that was removable only by their dentist. More recently the advantages of overdentures supported by two or more implants have been recognised. Both types of restorations have their place in the treatment of the compromised edentulous patient.

Bibliography

Textbooks

Brewer A A, Morrow R M. *Overdentures*. 2nd ed. St Louis: C V Mosby, 1980.

Grace A M, Smales F C. *Periodontal Control*. London: Quintessence, 1989.

Heartwell C M, Rahn A O. *Syllabus of complete dentures* (Chapter 24. Tooth supported complete denture). 2nd ed. Philadelphia: Lea and Febiger, 1974.

McDermott I G (ed). Removable partial overdentures. *Dent Clin North Am* 1990; 34.

Preiskel H W. *Precision attachments in prosthodontics: Overdentures and telescopic prostheses*. Vol. 2 Chicago: Quintessence, 1985.

Historical

Brill N. Adaptation and the hybrid prosthesis. *J Prosthet Dent* 1955; **5:** 811–824.

Ledger E. On preparing the mouth for the reception of a full set of artificial teeth. *Br J Dent Sci* 1856; **1:** 90.

Miller P A. Complete dentures supported by natural teeth. *J Prosthet Dent* 1958; **8:** 924–928.

Maintenance of bone

Attwood D A. Post extraction changes in the adult mandible as illustrated by microradiographs of midsagittal and serial cephalometric roentgenograms. *J Prosthet Dent* 1963; **13:** 680–824.

Crum R J, Rooney G E. Alveolar bone loss in overdentures: A five year study. *J Prosthet Dent* 1978; **40:** 610–613.

Tallgren A. The effect of denture wearing on facial morphology. *Acta Odontol Scand* 1967; **25:** 563–592.

Tallgren A. The continuing reduction of the residual alveolar ridges in complete denture wearers: A mixed-longitudinal study covering 25 years. *J Prosthet Dent* 1972; **27:** 120–132.

Sensory feedback

Adler P. Sensibility of teeth to loads applied in different directions. *J Dent Res* 1947; **26:** 279–289.

Bonaguro J G, Dusza G R, Bowman D C. Ability of human subjects to discriminate forces applied to certain teeth. *J Dent Res* 1969; **48:** 236–241.

Christensen L V, Levin A C. Periodontal discriminatory ability in human subjects with natural dentitions, overlay dentures and complete dentures. *J Dent Assoc S Afr* 1976; **31:** 339–342.

Crum R J, Loiselle R J. Oral perception and proprioception. A review of the literature and its significance to prosthodontics. *J Prosthet Dent* 1972; **28:** 215–230.

Hannan A G. Neuromuscular control of overdentures. In: *Precision attachments in prostho-*

dontics: Overdentures and telescopic prostheses. Vol. 2. Chicago: Quintessence, 1985, 21–27.

Haradlson T, Jemt T, Stalblad P-A, Lekholm U. Oral function in subjects with overdentures supported by osseointegrated implants. *Scand J Dent Res* 1988; **96**: 235–242.

Kay W D, Abes M S. Sensory perception in overdenture patients. *J Prosthet Dent* 1976; **35**: 615–619.

Loiselle R J, Crum R J, Rooney G E, Stuever C H. The physiologic basis for the overlay denture. *J Prosthet Dent* 1972; **28**: 4–12.

Manly R S, Pfaffman C, Lathrop D D, Keyser J. Oral sensory thresholds of persons with natural and artificial dentitions. *J Dent Res* 1952; **31**: 305–312.

Nagasawa T, Okane H, Tsuru H. The role of the periodontal ligament in overdenture treatment. *J Prosthet Dent* 1979; **42**: 12–16.

Pacer F J, Bowman D C. Occlusal force discrimination by denture patients. *J Prosthet Dent* 1975; **33**: 602–609.

Rissen L, House J E, Manly R S, Kapur K K. Clinical comparison of masticatory performance and electromyographic activity of patients with complete dentures, overdentures, and natural teeth. *J Prosthet Dent* 1978; **39**: 508–511.

Siirila H S, Laine P. The tactile sensitivity of the parodontium to slight axial loading of the teeth. *Acta Odontol Scand* 1963; **21**: 415–429.

Siirila H S, Laine P. Occlusal tactile threshold in denture wearers. *Acta Odontol Scand* 1969; **27**: 193–197.

Tryde G, Feydenberg D, Brill N. An assessment of the tactile sensitivity in human teeth. An evaluation of a quantitative method. *Acta Odontol Scand* 1962; **20**: 233–256.

Treatment planning and clinical techniques

Barsby M J. Partial dentures in the management of severe tooth wear. *Dental Update* 1988; **15**: 143–148.

Bates J F, Stafford G D. *Immediate complete dentures.* London: British Dental Association, 1971.

Brewer A A, Fenton A H. The overdenture. *Dent Clin North Am* 1973; **17**: 723–746.

Ettinger R L, Krell K. Endodontic problems in an overdenture population. *J Prosthet Dent* 1988; **59**: 459–462.

Fenton A H, Zarb G A, Mackay H F. Overdenture oversights. *Dent Clin North Am* 1979; **23**: 117–130.Langer Y, Langer A. Root-retained overdentures: Part II — Managing trauma between edentulous ridges and opposing dentition. *J Prosthet Dent* 1992; **67**: 77–81.

Lord J L, Teel S. The overdenture. *Dent Clin North Am* 1969; **13**: 871–881.

Lord J L, Teel S. The overdenture: Patient selection, use of copings and follow-up evaluation. *J Prosthet Dent* 1974; **32**: 41–51.

Lubow R M, Kretzchmar S L, Brown F H. The use of apically repositioned flaps in association with overdenture abutments. *Quintessence Int* 1988; **19**: 793–796.

Ralph J P, Murray F D. The use of root abutments in the support of complete dentures. *J Oral Rehabil* 1976; **3**: 293–297.

Ralph J P, Basker R M. The role of overdentures in gerodontics. *Dental Update* 1989; **16**: 353–360.

Richard G E, Sarka R J, Arnold R M, Knowles K I. Hemisected molars for additional overdenture support. *J Prosthet Dent* 1977; **38**: 16–21.

Schweitzer J M, Schweitzer R D, Schweitzer J. The telescoped complete denture: A research report at the clinical level. *J Prosthet Dent* 1971; **26**: 357–372.

Smith B G N. Toothwear: aetiology and diagnosis. *Dental Update* 1989; **16**: 204–212.

Smith B G N, Knight J F. An index for measuring the wear of teeth. *Br Dent J* 1984; **156**: 435–438.

Windchy A, Khan Z, Fields H. Overdentures with metal occlusion to maintain occlusal vertical dimension and prevent denture fracture. *J Prosthet Dent* 1988; **60**: 11–14.

Zamikoff I I. Overdentures — theory and technique. *J Am Dent Assoc* 1973; **86:** 853–857.

Attachments and magnets
Geissler P R. An inexpensive overdenture retainer. *Br Dent J* 1982; **153:** 194.
Gillings B R D. Magnetic denture retention systems: inexpensive and efficient. *Int Dent J* 1984; **34:** 184–197.
Gillings B R D. Magnetic retention for complete and partial overdentures. Part I. *J Prosthet Dent* 1981; **45:** 484–491.
Gillings B R D, Samant A. Overdentures with magnetic attachments. *Dent Clin North Am* 1990; **34:** 683–709.
Highton R, Caputo A A, Kinni A, Matyas J. The interaction of a magnetically retained denture with osseointegrated implants. *J Prosthet Dent* 60: 486–490.
Kurer P F. *The Kurer anchor system.* London: British Dental Association, 1980.
Laird W R E, Grant A A, Smith G A. The use of magnetic forces in prosthetic dentistry. *J Dent* 1981; **9:** 328–335.
Marquardt G L. Dolder bar-joint mandibular overdenture: A technique for nonparallel abutment teeth. *J Prosthet Dent* 1976; **36:** 101–111.
Mascola R F. The root-retained complete denture. *J Am Dent Assoc* 1976; **92:** 586–587.
Moghadam B K, Scandrett F R. Magnetic retention for overdentures. *J Prosthet Dent* 1979; **41:** 26–29.
Quinlivan J T. Fabrication of a simple ball-socket attachment. *J Prosthet Dent* 1974; **32:** 222–225.
Thayer H H, Caputo A A. Occlusal force transmission by overdenture attachments. *J Prosthet Dent* 1979; **41:** 266–271.
Warren A B, Caputo A A. Load transfer to alveolar bone as influenced by abutment designs for tooth-supported dentures. *J Prosthet Dent* 1975; **33:** 137–148.

Copy techniques
Duthie N, Lyon F F, Sturrock K C, Yemm R. A copying technique for replacement of complete dentures. *Br Dent J* 1978; **144:** 248–252.
Heath J R, Basker R M. The dimensional variability of duplicate dentures produced in an alginate investment. *Br Dent J* 1978; **144:** 111–114.
Heath J R, Johnson A. The versatility of the copy denture technique. *Br Dent J* 1981; **150:** 189–193.

Clinical evaluation and maintenance
Addy M. Chlorhexidine compared with other locally delivered antimicrobials. A short review. *J Clin Perio* 1986; **13:** 957–964.
Budtz-Jorgensen E. Effect of controlled oral hygiene in overdenture wearers: A 3-year study. *Int J Prosthodont* 1991; **4:** 226–231.
Budtz-Jørgensen E, Thylstrup A. The effect of controlled oral hygiene in overdenture wearers. *Acta Odontol Scand* 1988; **46:** 219–225.
Davis R K, Renner R P, Antos E W, Schlissel E R, Baer P N. A two-year longitudinal study of the periodontal health status of overdenture patients. *J Prosthet Dent* 1981; **45:** 358–363.
Dolder E J. The bar joint mandibular denture. *J Prosthet Dent* 1961; **1:** 689–707.
Ettinger R L, Taylor T D, Scandrett F R. Treatment needs of overdenture patients in a longitudinal study: Five-year results. *J Prosthet Dent* 1984; **52:** 532–537.
Ettinger R L. Tooth loss in an overdenture population. *J Prosthet Dent* 1988; **60:** 459–462.
Ettinger R L, Manderson D, Wefel J S, Jensen M E. An in-vitro evaluation of the prevention of caries on overdenture abutments. *J Dent Res* 1988; **67:** 1338–1341.
Ettinger R L, Jakobsen J. Caries: a problem in an overdenture population. *Community Dent Oral Epidemiol* 1990; **18:** 42–45.
Fenton A H, Hahn N. Tissue response to overdenture therapy. *J Prosthet Dent* 1978; **40:** 492–498.

Hussey D L, Linden G J. The efficacy of overdentures in clinical practice. *Br Dent J* 1986; **161:** 104–107.

Lofberg P G, Ericson G, Eliasson S. A clinical and radiographic examination of removable partial dentures retained by attachments to alveolar bars. *J Prosthet Dent* 1982; **47:** 126–132.

Lord J L, Teel S. The overdenture: Patient selection, use of copings and follow-up evaluation. *J Prosthet Dent* 1974; **32:** 41–51.

Mandel I D. Chemotherapeutic agents for controlling plaque and gingivitis. *J Clin Perio* 1988; **15:** 488–498.

McCord J F, Geissler P. A toothbrush for overdentures. *Dent Update* 1987; **14:** 447.

Rantenen T, Makila E, Yli-Urpo A, Siirila H S. Investigations of the therapeutic success with dentures retained by precision attachments.1. Root anchored complete overlay dentures. *Suon Hammaslaak Toim* 1971; **67:** 356–366.

Renner R P, Gomes B C, Shakun M L, Baer P N, Davis R K, Camp P. Four-year longitudinal study of the periodontal health status of overdenture patients. *J Prosthet Dent* 1984; **51:** 593–598.

Reitz P V, Weiner M G, Levin B. An overdenture survey: Preliminary report. *J Prosthet Dent* 1977; **37:** 246–258.

Scott J, Bates J F. The relining of partial dentures involving precision attachments. *J Prosthet Dent* 1972; **28:** 325–333.

Thayer H H, Caputo A A. Effects of overdentures upon remaining oral structures. *J Prosthet Dent* 1977; **37:** 374–381.

Toolson L B, Smith D E. A two-year longitudinal study of overdenture patients. Part I: Incidence and control of caries on overdenture abutments. *J Prosthet Dent* 1978; **40:** 486–491.

Toolson L B, Smith D E, Phillips C. A two-year longitudinal study of overdenture patients. Part II: Assessment of the periodontal health of overdenture abutments. *J Prosthet Dent* 1982; **47:** 4–11.

Toolson L B, Smith D E. A five-year longitudinal study of patients treated with overdentures. *J Prosthet Dent* 1983; **49:** 749–756.

Toolson L B, Taylor T D. A 10-year report of a longitudinal recall of overdenture patients. *J Prosthet Dent* 1989; **62:** 179–181.

Denture cleansers

Augsburger R H, Elahi J M. Evaluation of seven proprietary denture cleansers. *J Prosthet Dent* 1982; **47:** 356–359.

Budtz-Jorgensen E. Materials and methods for cleaning dentures. *J Prosthet Dent* 1979; **42:** 619–623.

Davenport J C, Wilson H J, Basker R M. The compatibility of tissue conditioners with denture cleaners and chlorhexidine. *J Dent* 1978; **6:** 239–246.

Ghalichebof M, Graser G N, Zander H A. The efficacy of denture-cleansing agents. *J Prosthet Dent* 1982; **48:** 515–520.

Harrison A, Basker R M, Smith I S. The compatibility of temporary soft materials with immersion denture cleansers. *Int J Prosthodont* 1989; **2:** 254–258.

Implant retained overdentures

Adell R, Eriksson B, Lekholm U, Brånemark P I, Jempt T. A long term follow-up study of osseointegrated implants in the treatment of totally edentulous jaws. *Int J Oral Maxillofac Implants* 1990; **5:** 347–352.

Adell R, Lekholm U, Rockler P, Brånemark P I. A fifteen year study of osseointegrated implants in the treatment of the edentulous jaw. *Int J Oral Surg* 1981; **10:** 387–416.

Albrektsson T. A multicenter report on osseointegrated oral implants. *J Prosthet Dent* 1988; **60:** 75–84.

Albrektsson T, Jacobsson M. Bone metal interface in osseointegration. *J Prosthet Dent* 1987; **57:** 597–607.

Albrektsson T, Zarb G A, Worthington P, Eriksson A R. The long term efficacy of currently used dental implants: A review and proposed criteria of success. *Int J Oral Maxillofac Implants* 1986; **1:** 11–25.

Andersson J E, Svartz K. CT Scanning in the preoperative planning of osseointegrated implants in the maxilla. *Int J Oral Maxillofac Surg* 1988; **17:** 33–35.

Armitage J E. Risk of blade implants. *In* Schnitman P, Shulman L (eds) *Dental Implants: Benefit and Risk.* pp 294-304. US Dept of Health and Human Services, NIH, 1980.

Bailey J H, Yanase R T, Bodine R L. The mandibular subperiosteal implant denture. A fourteen year study. *J Prosthet Dent* 1988; **60:** 358–361.

Beirne O R, Greenspan J S. Histologic evaluation of tissue response to hydroxylapatite implants in human mandibles. *J Dent Res* 1988; **64:** 1152–1154.

Brånemark P I, Hansson B O, Adell R, Breine U, Lindstrom J, Hallen O, Ohman A. Osseointegrated implants in the treatment of the edentulous jaw. Experience from a 10 year period. *Scand J Plast Reconstr Surg* 1977; **111(Suppl 16):** 1–132.

Bosker H. *The transmandibular implant.* PhD, University of Utrecht, 1986.

Boucher L T. Benefit and risk of subperiosteal implant: A critique. *In* Schnitman P, Shulman L (eds) *Dental Implants: Benefit and Risk.* pp 96-98. US Dept of Health and Human Services, NIH, 1980.

Engquist B, Bergendal T, Kallus T, Linden U. A retrospective multicenter evaluation of osseointegrated implants supporting overdentures. *Int J Oral Maxillofac Implants* 1988; **3:** 129–134.

Heimke G. Advanced ceramics for biomedical applications. *Adv Mater* 1989; **1:** 7–12.

Hjorting-Hansen E, Adawy A, Hillerup S. Mandibular vestibulolingual sulcoplasty with free skin graft. A five year follow up study. *J Oral Maxillofac Surg* 1983; **41:** 173–176.

Hopkins R. *A Colour Atlas of Preprosthetic Oral Surgery.* London: Wolfe Medical Publications Ltd, 1987.

Jempt T, Book K, Linden B, Urde G. Failures and complications in 92 consecutively inserted overdentures supported by Brånemark implants in severely resorbed edentulous maxillae: A study from prosthetic treatment to first annual check-up. *Int J Oral Maxillofac Implants* 1992; **7:** 162–167.

Jempt T, Stalblad P A. The effect of chewing movements on changing mandibular complete dentures to osseointegrated overdentures. *J Prosthet Dent* 1986; **55:** 357–361.

Johns R. Overdenture treatment with the Brånemark implant. *In* Albrektsson T, Zarb G A (eds). *The Brånemark Osseointegrated Implant.* pp 215–220. Chicago: Quintessence Publishing Co, 1989.

Johns R B, Jempt T, Heath M R, Hutton J E, McKenna S, McNamara D C, Van Steenberghe D, Taylor R, Watson R M, Herrmann I. A multi-centre study of overdentures supported by Brånemark implants. *Int J Oral Maxillofac Implants* 1992; **7:** 513–522.

Kapur K K. Benefit and risk of blade implants: A critique. *In* Schnitman P, Shulman L (eds) *Dental Implants: Benefit and Risk.* pp 306-308. US Dept of Health and Human Services, NIH, 1980.

Kent J N, Finger I M, Larsen H. Biointegrated hydroxylapatite coated dental implants: 5 year clinical observations. *J Am Dent Assoc* 1990; **121:** 138–144.

Lekholm U, Zarb G A. Patient selection and preparation. *In* Brånemark, P I, Zarb G A, Albrektsson T (eds). pp 199–209. *Tissue-Integrated Prostheses.* Chicago: Quintessence Publishing Co, 1985.

Linkow L I. Clinical evaluation of the various designed endosseous implants. *J Oral Implant Transplant Surg* 1966; **12:** 35–44.

Lindquist L W, Carlsson G E. Changes in masticatory function in complete denture wearers after insertion of bridges on osseointegrated implants in the lower jaw. *Adv Biomaterials* 1982; **4:** 151–155.

Lindquist L W, Carlsson G E, Glantz P O. Rehabilitation of the edentulous mandible with a tissue-integrated fixed prosthesis: A six-year longitudinal study. *Quintessence Int* 1987; **18:** 89–96.

Lindquist L W, Rockler B, Carlsson G E. Bone resorption around fixtures in edentulous patients treated with mandibular fixed tissue-integrated prostheses. *J Prosthet Dent* 1988; **59:** 59–63.

Quayle A A, Cawood J, Howell R A, Eldridge D J, Smith G A. The immediate or delayed replacement of teeth by permucosal intra-osseous implants: The Tubingen implant system: Part 1. Implant design, rationale for use and pre-operative assessment. *Br Dent J* 1989; **166:** 365–370.

Quayle A A, Cawood J, Smith G A, Eldridge D J, Howell R A. The immediate or delayed replacement of teeth by permucosal intra-osseous implants: The Tubingen implant system: Part 2. Surgical and restorative techniques. *Br Dent J* 1989; **166:** 403–410.

Reeve P E, Watson C J, Stafford G D. The role of personality in the management of complete denture patients. *Br Dent J* 1984; **156:** 356–362.

Small I A. The mandibular staple bone plate: Its use and role in prosthetic surgery. *J Head Neck Pathol* 1985; **4:** 111–116.

Small I A, Misiek D J. A sixteen year evaluation of the mandibular staple base plate. *J Oral Maxillofac Surg* 1986; **44:** 60–66.

Stalblad P A, Jansson T, Jempt T, Zarb G. Osseointegration in overdenture therapy. *Swed Dent J* 1985; Suppl 28: 169–170.

Stoelinga P J W, DeKoomen H A, Tideman H, Huijbers T J M. A reappraisal of the interposed bone graft augmentation of the atrophic mandible. *J Maxillofac Surg* 1983; **11:** 107–112.

Watson C J. Pressure changes at the denture base–mucosal surface interface resulting from mandibular vestibuloplasties. *Br Dent J* 1987; **163:** 11–18.

Wismeijer D, Vermeeren J I J H, Van Wass M A J. Patient satisfaction with overdentures supported by one stage TPS implants. *Int J Oral Maxillofac Implants* 1992; **7:** 51–55.

Index